Krav Maga Technique Logbook

Shawn Kovacich

AUTHOR APPROVED SHOOT/NO SHOOT EDITION

Printed in the United States of America

Library of Congress Control Number: 2019912348

ISBN #978-1-7336809-5-0

Table of Contents

Introduction

Disclaimer....vii
Safety....viii
And Finally....viii
How To Use This Logbook....ix
Cover and Spine....ix
Logbook Entry....ix
Technique....ix
L - R - Alt....ix
Start Date....ix
Notes....x
Slow Reps....x
Reg Reps....x
Notes....x
End Date....xi
Code Key....xi
Code Key....xii

Logbook Entry (write down the name of the technique you are practicing)

________ 1
________ 3
________ 5
________ 7
________ 9
________ 11
________ 13
________ 15
________ 17
________ 19
________ 21
________ 23
________ 25
________ 27
________ 29
________ 31

Logbook Entry (continued)

______ 33
______ 35
______ 37
______ 39
______ 41
______ 43
______ 45
______ 47
______ 49
______ 51
______ 53
______ 55
______ 57
______ 59
______ 61
______ 63
______ 65
______ 67
______ 69
______ 71
______ 73
______ 75
______ 77
______ 79
______ 81
______ 83
______ 85
______ 87
______ 89
______ 91
______ 93
______ 95
______ 97
______ 99

Additional Notes..101

Disclaimer

Please note that the author, Shawn Kovacich, and the publisher, **Shoot-NO-Shoot Productions**, Chikara Kan, Inc., of this logbook are **Not Responsible** in any manner whatsoever, for any injury, which may occur by reading and/or following the instructions located within this logbook. The distributor and any and all other individuals and or businesses connected with the production and/or distribution of this book also disclaim any and all liability for any damage or injury of any kind that may result from the appropriate use or inappropriate misuse of any information contained within this book.

The techniques that you will be executing while utilizing this logbook are potentially lethal in nature and are intended to be used in a **safe and law abiding manner** during practice and also if/when confronted with a potentially life threatening situation that may or may not require the use of lethal force in order to protect ones life. It is the reader or readers direct responsibility to research and comply with all applicable laws regarding the use of force in a self-defense situation in the areas in which they reside, do business, or frequent. **Ignorance of the law never has been, nor is it likely that it ever will be, a viable defense.**

Due to the fact that the techniques described in this book are physical in nature, it is essential that the reader or readers of this book consult a qualified and competent physician before following any of the activities, physical or otherwise, which are described within this book.

This book and all the information contained therein is intended to be used as a supplemental training aid, and should be used as such, under the guidance of a qualified and competent martial arts/self-defense instructor.

This book and the information contained within are for informational purposes only!

Copyrights & Trademarks

SAFETY

Due to the physical nature of the activities that you will be participating in while utilizing this logbook, it is highly recommended that you first consult a qualified and competent physician before starting any physical training routine. Regardless of your current physical condition. Remember, an ounce of prevention is worth a pound of cure. And also remember, **Safety First, Foremost, and Always!**

AND FINALLY...

You would do well to remember that progress should be measured in inches, not in miles. You are not going to execute 300 or 400 repetitions of a particular technique and then all of a sudden be showing a remarkable improvement in your self-defensive skills. It takes time, proper practice, full immersion and active participation, and consistency in practice in order to achieve positive results. You probably won't see any big changes after just a few hundred repetitions, but you should be seeing some significant improvement after the first 1,000 repetitions. And, you should continue to show improvement after every 1,000 repetitions of a particular technique that you execute, if you have fully immersed yourself both physically and mentally, while actively participating in each and every training session.

In addition to that, if you have also been studying and refining your self-defensive skills under the watchful eye of a qualified and competent martial arts/self-defense instructor, then you will make even greater progress in honing your self-defensive skills.

HOW TO USE THIS LOGBOOK

In order to obtain the full benefits of this **Krav Maga: Technique Logbook**, you will want to follow these simple instructions on how to best utilize this logbook. This logbook should be used along with your active participation with the various techniques that you are currently practicing. These techniques that you have learned and/or are learning, should be practiced on a regular and consistent basis while keeping an extensive and detailed record of the number of repetitions that your are executing for each particular technique.

However, please make sure that you are first doing the techniques correctly before you start keeping track of the thousands of repetitions that you are going to start keeping track of. Because if you are doing the techniques incorrectly the first time, and you continue to practice them incorrectly, how do you think you are going to execute them the 6,000th time?

COVER & SPINE

When you look at the front cover and the spine, you will see a small white box in the lower left hand corner on the front cover and a smaller white box on the spine. These are include in order to give you a space in which to write down the volume number or other such indicator of the logbook that you are currently using.

LOGBOOK ENTRY (on the **"Table of Contents"** pages)

This is where you will write down the name of the particular **"Technique"** that you are currently practicing.

TECHNIQUE

Write down the name of the **"Technique"** that you are currently practicing and keeping meticulous track of. For example; Straight Punch, Forward Elbow, Knee, Front Kick, etc.

LEFT - RIGHT - ALT

This is where you will enter whether you are exclusively using the **"Left"** arm/leg/side, or the **"Right"** arm/leg/side for the execution of this particular **"Technique."** Or are you alternating between the left and right (**"Alt"**).

START DATE

This section allows you to write down any particular **"Notes"** that you may have concerning this particular **"Technique"** that you are keeping track of.

NOTES

This section allows you to write down any particular **"Notes"** that you may have concerning the execution of this particular **"Technique"** that you are keeping track of.

SLOW REPS

This section allows you to keep track of the number of slow repetitions or **"Slow Reps"** that you execute throughout the time frame that you are also executing your regular repetitions of this particular technique. My personal recommendation is that you should execute at least 10 **"Slow Reps"** at the beginning and at the end of every session that you are working on this particular technique.

For example if you are going to perform 100 **Straight Punches** alternating **"Alt"** hands by executing 50 punches with each hand. You should start out be executing 10 slow repetitions with each hand followed by 50 repetitions with each hand at normal speed and then finish up with another 10 slow repetitions with each hand. That would give you a total of 140 punches executed in that particular training session, 20 **"Slow Reps"** with each hand and 50 **"Reg Reps"** with each hand.

REG REPS

The regular repetitions or **"Reg Reps"** section allows you to circle or check off the number of regular repetitions of this particular technique that you are keeping track of in increments of 10. So, for every 10 repetitions that you execute of a particular technique, you would check off the next 10 repetition increment. For example; if you are just starting out and you just finished executing say 50 repetitions of **Straight Punches** with your **"Left"** hand, you would then circle or check off the following numbers; 10, 20, 30, 40, and 50.

However, if you are executing punches alternating **"Alt"** with both hands, you would need to make sure that you are executing an equal number of punches with each hand so that when divided by 2 it would give you a number ending in zero. For example; if you executed a total of 70 punches (35 with each hand) that wouldn't be divisible by two in order to come up with a number ending in zero. So you would need to execute 5 more punches with each hand in order to come up with a total of 80 punches which you would then divide by 2 in order to come up with a number ending in zero, which would be 40 punches executed with each hand. You would then circle or check off the following numbers (if you started out at zero); 10, 20, 30, and 40.

NOTES

This **"Notes"** section, allows you the opportunity to write down any particular notes that you may have concerning this particular **"Technique"** after every 1,000 repetitions executed. There is no notes section after the **"Slow Reps,"** as your slow repetitions are intended to be executed at various stages during the execution of the 5,000 regular repetitions or "**Reg Reps.**" Be sure and include the month, day, and year. Use the **"Additional Notes"** section included at the end of this logbook to keep track of any additional items that you feel you need to keep track of during the course of utilizing this logbook.

END DATE

This is where you will enter the **"Date"** in which you completed the 1,000 slow repetitions **"Slow Reps"** and the 5,000 regular repetitions **"Reg Reps"** of this particular **"Technique."**

CODE KEY

This section was included in order to allow you the opportunity to personalize the way you keep track of your progress by allowing you to use a variety of symbols, ink colors, etc., in order to keep track of any particularities that you may want to include in your own personal way to add to your training regiment. I have included some various possibilities that you could use, or come up with your own that may fit your own particular needs. So without further ado, let's get started.

BLACK INK = Physically and mentally an average training routine.

BLUE INK = Physically and mentally not feeling well. Not a typical training routine.

RED INK = Really pumped up both physically and mentally. Outstanding training routine.

X (over the number) = Executed that particular number of repetitions while doing line drills, etc.

CIRCLE (around the number) = Executed that particular number of repetitions while executing the technique in the air.

SQUARE (around the number) = Executed that particular number of repetitions while striking an object, ie; heavy bag, kicking "Shield," focus mitts, etc.

TRIANGLE (around the number) = Executed that particular number of repetitions while using a training partner as a target.

CODE KEY

Here is the **"Code Key"** for you to personalize to your own particular needs and training regiment. I would strongly recommend that you read through the **"How To Use This Logbook"** section prior to using this logbook in order to maximize its use to its fullest potential. Be sure and fill this out prior to starting your logbook.

BLACK INK ______________________________

____

BLUE INK ______________________________

____

RED INK ______________________________

____

X ______________________________

____

CIRCLE ______________________________

____

SQUARE ______________________________

____

TRIANGLE ______________________________

____

Technique:______________________ **L - R - Alt. Start Date:** _______
Notes:____

____

____

____

Slow Reps: 10 - 20 - 30 - 40 - 50 - 60 - 70 - 80 - 90 - 100 - 110 - 120 - 130
140 - 150 - 160 - 170 - 180 - 190 - 200 - 210 - 220 - 230 - 240 - 250 - 260
270 - 280 - 290 - 300 - 310 - 320 - 330 - 340 - 350 - 360 - 370 - 380 - 390
400 - 410 - 420 - 430 - 440 - 450 - 460 - 470 - 480 - 490 - 500 - 510 - 520
530 - 540 - 550 - 560 - 570 - 580 - 590 - 600 - 610 - 620 - 630 - 640 - 650
660 - 670 - 680 - 690 - 700 - 710 - 720 - 730 - 740 - 750 - 760 - 770 - 780
790 - 800 - 810 - 820 - 830 - 840 - 850 - 860 - 870 - 880 - 890 - 900 - 910
920 - 930 - 940 - 950 - 960 - 970 - 980 - 990 - 1000

Reg Reps: 10 - 20 - 30 - 40 - 50 - 60 - 70 - 80 - 90 - 100 - 110 - 120 - 130
140 - 150 - 160 - 170 - 180 - 190 - 200 - 210 - 220 - 230 - 240 - 250 - 260
270 - 280 - 290 - 300 - 310 - 320 - 330 - 340 - 350 - 360 - 370 - 380 - 390
400 - 410 - 420 - 430 - 440 - 450 - 460 - 470 - 480 - 490 - 500 - 510 - 520
530 - 540 - 550 - 560 - 570 - 580 - 590 - 600 - 610 - 620 - 630 - 640 - 650
660 - 670 - 680 - 690 - 700 - 710 - 720 - 730 - 740 - 750 - 760 - 770 - 780
790 - 800 - 810 - 820 - 830 - 840 - 850 - 860 - 870 - 880 - 890 - 900 - 910
920 - 930 - 940 - 950 - 960 - 970 - 980 - 990 - 1000
Notes: ____

____

____

____

Reg Reps: 10 - 20 - 30 - 40 - 50 - 60 - 70 - 80 - 90 - 100 - 110 - 120 - 130
140 - 150 - 160 - 170 - 180 - 190 - 200 - 210 - 220 - 230 - 240 - 250 - 260
270 - 280 - 290 - 300 - 310 - 320 - 330 - 340 - 350 - 360 - 370 - 380 - 390
400 - 410 - 420 - 430 - 440 - 450 - 460 - 470 - 480 - 490 - 500 - 510 - 520
530 - 540 - 550 - 560 - 570 - 580 - 590 - 600 - 610 - 620 - 630 - 640 - 650
660 - 670 - 680 - 690 - 700 - 710 - 720 - 730 - 740 - 750 - 760 - 770 - 780
790 - 800 - 810 - 820 - 830 - 840 - 850 - 860 - 870 - 880 - 890 - 900 - 910
920 - 930 - 940 - 950 - 960 - 970 - 980 - 990 - 1000
Notes: ____

____

____

____

Reg Reps: 10 - 20 - 30 - 40 - 50 - 60 - 70 - 80 - 90 - 100 - 110 - 120 - 130
140 - 150 - 160 - 170 - 180 - 190 - 200 - 210 - 220 - 230 - 240 - 250 - 260
270 - 280 - 290 - 300 - 310 - 320 - 330 - 340 - 350 - 360 - 370 - 380 - 390
400 - 410 - 420 - 430 - 440 - 450 - 460 - 470 - 480 - 490 - 500 - 510 - 520
530 - 540 - 550 - 560 - 570 - 580 - 590 - 600 - 610 - 620 - 630 - 640 - 650
660 - 670 - 680 - 690 - 700 - 710 - 720 - 730 - 740 - 750 - 760 - 770 - 780
790 - 800 - 810 - 820 - 830 - 840 - 850 - 860 - 870 - 880 - 890 - 900 - 910
920 - 930 - 940 - 950 - 960 - 970 - 980 - 990 - 1000

Notes: ______________________________

____

____

____

Reg Reps: 10 - 20 - 30 - 40 - 50 - 60 - 70 - 80 - 90 - 100 - 110 - 120 - 130
140 - 150 - 160 - 170 - 180 - 190 - 200 - 210 - 220 - 230 - 240 - 250 - 260
270 - 280 - 290 - 300 - 310 - 320 - 330 - 340 - 350 - 360 - 370 - 380 - 390
400 - 410 - 420 - 430 - 440 - 450 - 460 - 470 - 480 - 490 - 500 - 510 - 520
530 - 540 - 550 - 560 - 570 - 580 - 590 - 600 - 610 - 620 - 630 - 640 - 650
660 - 670 - 680 - 690 - 700 - 710 - 720 - 730 - 740 - 750 - 760 - 770 - 780
790 - 800 - 810 - 820 - 830 - 840 - 850 - 860 - 870 - 880 - 890 - 900 - 910
920 - 930 - 940 - 950 - 960 - 970 - 980 - 990 - 1000

Notes: ______________________________

____

____

____

Reg Reps: 10 - 20 - 30 - 40 - 50 - 60 - 70 - 80 - 90 - 100 - 110 - 120 - 130
140 - 150 - 160 - 170 - 180 - 190 - 200 - 210 - 220 - 230 - 240 - 250 - 260
270 - 280 - 290 - 300 - 310 - 320 - 330 - 340 - 350 - 360 - 370 - 380 - 390
400 - 410 - 420 - 430 - 440 - 450 - 460 - 470 - 480 - 490 - 500 - 510 - 520
530 - 540 - 550 - 560 - 570 - 580 - 590 - 600 - 610 - 620 - 630 - 640 - 650
660 - 670 - 680 - 690 - 700 - 710 - 720 - 730 - 740 - 750 - 760 - 770 - 780
790 - 800 - 810 - 820 - 830 - 840 - 850 - 860 - 870 - 880 - 890 - 900 - 910
920 - 930 - 940 - 950 - 960 - 970 - 980 - 990 - 1000

Notes: ______________________________

____

____

______________________________ **End Date:** __________

Technique:________________________ **L - R - Alt. Start Date:** _______
Notes:____

Slow Reps: 10 - 20 - 30 - 40 - 50 - 60 - 70 - 80 - 90 - 100 - 110 - 120 - 130
140 - 150 - 160 - 170 - 180 - 190 - 200 - 210 - 220 - 230 - 240 - 250 - 260
270 - 280 - 290 - 300 - 310 - 320 - 330 - 340 - 350 - 360 - 370 - 380 - 390
400 - 410 - 420 - 430 - 440 - 450 - 460 - 470 - 480 - 490 - 500 - 510 - 520
530 - 540 - 550 - 560 - 570 - 580 - 590 - 600 - 610 - 620 - 630 - 640 - 650
660 - 670 - 680 - 690 - 700 - 710 - 720 - 730 - 740 - 750 - 760 - 770 - 780
790 - 800 - 810 - 820 - 830 - 840 - 850 - 860 - 870 - 880 - 890 - 900 - 910
920 - 930 - 940 - 950 - 960 - 970 - 980 - 990 - 1000

Reg Reps: 10 - 20 - 30 - 40 - 50 - 60 - 70 - 80 - 90 - 100 - 110 - 120 - 130
140 - 150 - 160 - 170 - 180 - 190 - 200 - 210 - 220 - 230 - 240 - 250 - 260
270 - 280 - 290 - 300 - 310 - 320 - 330 - 340 - 350 - 360 - 370 - 380 - 390
400 - 410 - 420 - 430 - 440 - 450 - 460 - 470 - 480 - 490 - 500 - 510 - 520
530 - 540 - 550 - 560 - 570 - 580 - 590 - 600 - 610 - 620 - 630 - 640 - 650
660 - 670 - 680 - 690 - 700 - 710 - 720 - 730 - 740 - 750 - 760 - 770 - 780
790 - 800 - 810 - 820 - 830 - 840 - 850 - 860 - 870 - 880 - 890 - 900 - 910
920 - 930 - 940 - 950 - 960 - 970 - 980 - 990 - 1000
Notes: ____

Reg Reps: 10 - 20 - 30 - 40 - 50 - 60 - 70 - 80 - 90 - 100 - 110 - 120 - 130
140 - 150 - 160 - 170 - 180 - 190 - 200 - 210 - 220 - 230 - 240 - 250 - 260
270 - 280 - 290 - 300 - 310 - 320 - 330 - 340 - 350 - 360 - 370 - 380 - 390
400 - 410 - 420 - 430 - 440 - 450 - 460 - 470 - 480 - 490 - 500 - 510 - 520
530 - 540 - 550 - 560 - 570 - 580 - 590 - 600 - 610 - 620 - 630 - 640 - 650
660 - 670 - 680 - 690 - 700 - 710 - 720 - 730 - 740 - 750 - 760 - 770 - 780
790 - 800 - 810 - 820 - 830 - 840 - 850 - 860 - 870 - 880 - 890 - 900 - 910
920 - 930 - 940 - 950 - 960 - 970 - 980 - 990 - 1000
Notes: ____

Reg Reps: 10 - 20 - 30 - 40 - 50 - 60 - 70 - 80 - 90 - 100 - 110 - 120 - 130
140 - 150 - 160 - 170 - 180 - 190 - 200 - 210 - 220 - 230 - 240 - 250 - 260
270 - 280 - 290 - 300 - 310 - 320 - 330 - 340 - 350 - 360 - 370 - 380 - 390
400 - 410 - 420 - 430 - 440 - 450 - 460 - 470 - 480 - 490 - 500 - 510 - 520
530 - 540 - 550 - 560 - 570 - 580 - 590 - 600 - 610 - 620 - 630 - 640 - 650
660 - 670 - 680 - 690 - 700 - 710 - 720 - 730 - 740 - 750 - 760 - 770 - 780
790 - 800 - 810 - 820 - 830 - 840 - 850 - 860 - 870 - 880 - 890 - 900 - 910
920 - 930 - 940 - 950 - 960 - 970 - 980 - 990 - 1000

Notes: ____________________

____

____

____

Reg Reps: 10 - 20 - 30 - 40 - 50 - 60 - 70 - 80 - 90 - 100 - 110 - 120 - 130
140 - 150 - 160 - 170 - 180 - 190 - 200 - 210 - 220 - 230 - 240 - 250 - 260
270 - 280 - 290 - 300 - 310 - 320 - 330 - 340 - 350 - 360 - 370 - 380 - 390
400 - 410 - 420 - 430 - 440 - 450 - 460 - 470 - 480 - 490 - 500 - 510 - 520
530 - 540 - 550 - 560 - 570 - 580 - 590 - 600 - 610 - 620 - 630 - 640 - 650
660 - 670 - 680 - 690 - 700 - 710 - 720 - 730 - 740 - 750 - 760 - 770 - 780
790 - 800 - 810 - 820 - 830 - 840 - 850 - 860 - 870 - 880 - 890 - 900 - 910
920 - 930 - 940 - 950 - 960 - 970 - 980 - 990 - 1000

Notes: ____________________

____

____

____

Reg Reps: 10 - 20 - 30 - 40 - 50 - 60 - 70 - 80 - 90 - 100 - 110 - 120 - 130
140 - 150 - 160 - 170 - 180 - 190 - 200 - 210 - 220 - 230 - 240 - 250 - 260
270 - 280 - 290 - 300 - 310 - 320 - 330 - 340 - 350 - 360 - 370 - 380 - 390
400 - 410 - 420 - 430 - 440 - 450 - 460 - 470 - 480 - 490 - 500 - 510 - 520
530 - 540 - 550 - 560 - 570 - 580 - 590 - 600 - 610 - 620 - 630 - 640 - 650
660 - 670 - 680 - 690 - 700 - 710 - 720 - 730 - 740 - 750 - 760 - 770 - 780
790 - 800 - 810 - 820 - 830 - 840 - 850 - 860 - 870 - 880 - 890 - 900 - 910
920 - 930 - 940 - 950 - 960 - 970 - 980 - 990 - 1000

Notes: ____________________

____

____

____________________ **End Date:** __________

Technique:____________________ **L - R - Alt. Start Date:** _______

Notes:

Slow Reps: 10 - 20 - 30 - 40 - 50 - 60 - 70 - 80 - 90 - 100 - 110 - 120 - 130
140 - 150 - 160 - 170 - 180 - 190 - 200 - 210 - 220 - 230 - 240 - 250 - 260
270 - 280 - 290 - 300 - 310 - 320 - 330 - 340 - 350 - 360 - 370 - 380 - 390
400 - 410 - 420 - 430 - 440 - 450 - 460 - 470 - 480 - 490 - 500 - 510 - 520
530 - 540 - 550 - 560 - 570 - 580 - 590 - 600 - 610 - 620 - 630 - 640 - 650
660 - 670 - 680 - 690 - 700 - 710 - 720 - 730 - 740 - 750 - 760 - 770 - 780
790 - 800 - 810 - 820 - 830 - 840 - 850 - 860 - 870 - 880 - 890 - 900 - 910
920 - 930 - 940 - 950 - 960 - 970 - 980 - 990 - 1000

Reg Reps: 10 - 20 - 30 - 40 - 50 - 60 - 70 - 80 - 90 - 100 - 110 - 120 - 130
140 - 150 - 160 - 170 - 180 - 190 - 200 - 210 - 220 - 230 - 240 - 250 - 260
270 - 280 - 290 - 300 - 310 - 320 - 330 - 340 - 350 - 360 - 370 - 380 - 390
400 - 410 - 420 - 430 - 440 - 450 - 460 - 470 - 480 - 490 - 500 - 510 - 520
530 - 540 - 550 - 560 - 570 - 580 - 590 - 600 - 610 - 620 - 630 - 640 - 650
660 - 670 - 680 - 690 - 700 - 710 - 720 - 730 - 740 - 750 - 760 - 770 - 780
790 - 800 - 810 - 820 - 830 - 840 - 850 - 860 - 870 - 880 - 890 - 900 - 910
920 - 930 - 940 - 950 - 960 - 970 - 980 - 990 - 1000

Notes:

Reg Reps: 10 - 20 - 30 - 40 - 50 - 60 - 70 - 80 - 90 - 100 - 110 - 120 - 130
140 - 150 - 160 - 170 - 180 - 190 - 200 - 210 - 220 - 230 - 240 - 250 - 260
270 - 280 - 290 - 300 - 310 - 320 - 330 - 340 - 350 - 360 - 370 - 380 - 390
400 - 410 - 420 - 430 - 440 - 450 - 460 - 470 - 480 - 490 - 500 - 510 - 520
530 - 540 - 550 - 560 - 570 - 580 - 590 - 600 - 610 - 620 - 630 - 640 - 650
660 - 670 - 680 - 690 - 700 - 710 - 720 - 730 - 740 - 750 - 760 - 770 - 780
790 - 800 - 810 - 820 - 830 - 840 - 850 - 860 - 870 - 880 - 890 - 900 - 910
920 - 930 - 940 - 950 - 960 - 970 - 980 - 990 - 1000

Notes:

Reg Reps: 10 - 20 - 30 - 40 - 50 - 60 - 70 - 80 - 90 - 100 - 110 - 120 - 130
140 - 150 - 160 - 170 - 180 - 190 - 200 - 210 - 220 - 230 - 240 - 250 - 260
270 - 280 - 290 - 300 - 310 - 320 - 330 - 340 - 350 - 360 - 370 - 380 - 390
400 - 410 - 420 - 430 - 440 - 450 - 460 - 470 - 480 - 490 - 500 - 510 - 520
530 - 540 - 550 - 560 - 570 - 580 - 590 - 600 - 610 - 620 - 630 - 640 - 650
660 - 670 - 680 - 690 - 700 - 710 - 720 - 730 - 740 - 750 - 760 - 770 - 780
790 - 800 - 810 - 820 - 830 - 840 - 850 - 860 - 870 - 880 - 890 - 900 - 910
920 - 930 - 940 - 950 - 960 - 970 - 980 - 990 - 1000

Notes: ____

____

____

____

Reg Reps: 10 - 20 - 30 - 40 - 50 - 60 - 70 - 80 - 90 - 100 - 110 - 120 - 130
140 - 150 - 160 - 170 - 180 - 190 - 200 - 210 - 220 - 230 - 240 - 250 - 260
270 - 280 - 290 - 300 - 310 - 320 - 330 - 340 - 350 - 360 - 370 - 380 - 390
400 - 410 - 420 - 430 - 440 - 450 - 460 - 470 - 480 - 490 - 500 - 510 - 520
530 - 540 - 550 - 560 - 570 - 580 - 590 - 600 - 610 - 620 - 630 - 640 - 650
660 - 670 - 680 - 690 - 700 - 710 - 720 - 730 - 740 - 750 - 760 - 770 - 780
790 - 800 - 810 - 820 - 830 - 840 - 850 - 860 - 870 - 880 - 890 - 900 - 910
920 - 930 - 940 - 950 - 960 - 970 - 980 - 990 - 1000

Notes: ____

____

____

____

Reg Reps: 10 - 20 - 30 - 40 - 50 - 60 - 70 - 80 - 90 - 100 - 110 - 120 - 130
140 - 150 - 160 - 170 - 180 - 190 - 200 - 210 - 220 - 230 - 240 - 250 - 260
270 - 280 - 290 - 300 - 310 - 320 - 330 - 340 - 350 - 360 - 370 - 380 - 390
400 - 410 - 420 - 430 - 440 - 450 - 460 - 470 - 480 - 490 - 500 - 510 - 520
530 - 540 - 550 - 560 - 570 - 580 - 590 - 600 - 610 - 620 - 630 - 640 - 650
660 - 670 - 680 - 690 - 700 - 710 - 720 - 730 - 740 - 750 - 760 - 770 - 780
790 - 800 - 810 - 820 - 830 - 840 - 850 - 860 - 870 - 880 - 890 - 900 - 910
920 - 930 - 940 - 950 - 960 - 970 - 980 - 990 - 1000

Notes: ____

____

____

________________________________ **End Date:** ____________

Technique:________________________ **L - R - Alt. Start Date:** _______

Notes:____

Slow Reps: 10 - 20 - 30 - 40 - 50 - 60 - 70 - 80 - 90 - 100 - 110 - 120 - 130
140 - 150 - 160 - 170 - 180 - 190 - 200 - 210 - 220 - 230 - 240 - 250 - 260
270 - 280 - 290 - 300 - 310 - 320 - 330 - 340 - 350 - 360 - 370 - 380 - 390
400 - 410 - 420 - 430 - 440 - 450 - 460 - 470 - 480 - 490 - 500 - 510 - 520
530 - 540 - 550 - 560 - 570 - 580 - 590 - 600 - 610 - 620 - 630 - 640 - 650
660 - 670 - 680 - 690 - 700 - 710 - 720 - 730 - 740 - 750 - 760 - 770 - 780
790 - 800 - 810 - 820 - 830 - 840 - 850 - 860 - 870 - 880 - 890 - 900 - 910
920 - 930 - 940 - 950 - 960 - 970 - 980 - 990 - 1000

Reg Reps: 10 - 20 - 30 - 40 - 50 - 60 - 70 - 80 - 90 - 100 - 110 - 120 - 130
140 - 150 - 160 - 170 - 180 - 190 - 200 - 210 - 220 - 230 - 240 - 250 - 260
270 - 280 - 290 - 300 - 310 - 320 - 330 - 340 - 350 - 360 - 370 - 380 - 390
400 - 410 - 420 - 430 - 440 - 450 - 460 - 470 - 480 - 490 - 500 - 510 - 520
530 - 540 - 550 - 560 - 570 - 580 - 590 - 600 - 610 - 620 - 630 - 640 - 650
660 - 670 - 680 - 690 - 700 - 710 - 720 - 730 - 740 - 750 - 760 - 770 - 780
790 - 800 - 810 - 820 - 830 - 840 - 850 - 860 - 870 - 880 - 890 - 900 - 910
920 - 930 - 940 - 950 - 960 - 970 - 980 - 990 - 1000

Notes: ____

Reg Reps: 10 - 20 - 30 - 40 - 50 - 60 - 70 - 80 - 90 - 100 - 110 - 120 - 130
140 - 150 - 160 - 170 - 180 - 190 - 200 - 210 - 220 - 230 - 240 - 250 - 260
270 - 280 - 290 - 300 - 310 - 320 - 330 - 340 - 350 - 360 - 370 - 380 - 390
400 - 410 - 420 - 430 - 440 - 450 - 460 - 470 - 480 - 490 - 500 - 510 - 520
530 - 540 - 550 - 560 - 570 - 580 - 590 - 600 - 610 - 620 - 630 - 640 - 650
660 - 670 - 680 - 690 - 700 - 710 - 720 - 730 - 740 - 750 - 760 - 770 - 780
790 - 800 - 810 - 820 - 830 - 840 - 850 - 860 - 870 - 880 - 890 - 900 - 910
920 - 930 - 940 - 950 - 960 - 970 - 980 - 990 - 1000

Notes: ____

Reg Reps: 10 - 20 - 30 - 40 - 50 - 60 - 70 - 80 - 90 - 100 - 110 - 120 - 130
140 - 150 - 160 - 170 - 180 - 190 - 200 - 210 - 220 - 230 - 240 - 250 - 260
270 - 280 - 290 - 300 - 310 - 320 - 330 - 340 - 350 - 360 - 370 - 380 - 390
400 - 410 - 420 - 430 - 440 - 450 - 460 - 470 - 480 - 490 - 500 - 510 - 520
530 - 540 - 550 - 560 - 570 - 580 - 590 - 600 - 610 - 620 - 630 - 640 - 650
660 - 670 - 680 - 690 - 700 - 710 - 720 - 730 - 740 - 750 - 760 - 770 - 780
790 - 800 - 810 - 820 - 830 - 840 - 850 - 860 - 870 - 880 - 890 - 900 - 910
920 - 930 - 940 - 950 - 960 - 970 - 980 - 990 - 1000

Notes: ____

____

____

____

Reg Reps: 10 - 20 - 30 - 40 - 50 - 60 - 70 - 80 - 90 - 100 - 110 - 120 - 130
140 - 150 - 160 - 170 - 180 - 190 - 200 - 210 - 220 - 230 - 240 - 250 - 260
270 - 280 - 290 - 300 - 310 - 320 - 330 - 340 - 350 - 360 - 370 - 380 - 390
400 - 410 - 420 - 430 - 440 - 450 - 460 - 470 - 480 - 490 - 500 - 510 - 520
530 - 540 - 550 - 560 - 570 - 580 - 590 - 600 - 610 - 620 - 630 - 640 - 650
660 - 670 - 680 - 690 - 700 - 710 - 720 - 730 - 740 - 750 - 760 - 770 - 780
790 - 800 - 810 - 820 - 830 - 840 - 850 - 860 - 870 - 880 - 890 - 900 - 910
920 - 930 - 940 - 950 - 960 - 970 - 980 - 990 - 1000

Notes: ____

____

____

____

Reg Reps: 10 - 20 - 30 - 40 - 50 - 60 - 70 - 80 - 90 - 100 - 110 - 120 - 130
140 - 150 - 160 - 170 - 180 - 190 - 200 - 210 - 220 - 230 - 240 - 250 - 260
270 - 280 - 290 - 300 - 310 - 320 - 330 - 340 - 350 - 360 - 370 - 380 - 390
400 - 410 - 420 - 430 - 440 - 450 - 460 - 470 - 480 - 490 - 500 - 510 - 520
530 - 540 - 550 - 560 - 570 - 580 - 590 - 600 - 610 - 620 - 630 - 640 - 650
660 - 670 - 680 - 690 - 700 - 710 - 720 - 730 - 740 - 750 - 760 - 770 - 780
790 - 800 - 810 - 820 - 830 - 840 - 850 - 860 - 870 - 880 - 890 - 900 - 910
920 - 930 - 940 - 950 - 960 - 970 - 980 - 990 - 1000

Notes: ____

____

____

______________________________________ **End Date:** __________

Technique:______________________ **L - R - Alt. Start Date:** _______

Notes:____

____

____

____

Slow Reps: 10 - 20 - 30 - 40 - 50 - 60 - 70 - 80 - 90 - 100 - 110 - 120 - 130
140 - 150 - 160 - 170 - 180 - 190 - 200 - 210 - 220 - 230 - 240 - 250 - 260
270 - 280 - 290 - 300 - 310 - 320 - 330 - 340 - 350 - 360 - 370 - 380 - 390
400 - 410 - 420 - 430 - 440 - 450 - 460 - 470 - 480 - 490 - 500 - 510 - 520
530 - 540 - 550 - 560 - 570 - 580 - 590 - 600 - 610 - 620 - 630 - 640 - 650
660 - 670 - 680 - 690 - 700 - 710 - 720 - 730 - 740 - 750 - 760 - 770 - 780
790 - 800 - 810 - 820 - 830 - 840 - 850 - 860 - 870 - 880 - 890 - 900 - 910
920 - 930 - 940 - 950 - 960 - 970 - 980 - 990 - 1000

Reg Reps: 10 - 20 - 30 - 40 - 50 - 60 - 70 - 80 - 90 - 100 - 110 - 120 - 130
140 - 150 - 160 - 170 - 180 - 190 - 200 - 210 - 220 - 230 - 240 - 250 - 260
270 - 280 - 290 - 300 - 310 - 320 - 330 - 340 - 350 - 360 - 370 - 380 - 390
400 - 410 - 420 - 430 - 440 - 450 - 460 - 470 - 480 - 490 - 500 - 510 - 520
530 - 540 - 550 - 560 - 570 - 580 - 590 - 600 - 610 - 620 - 630 - 640 - 650
660 - 670 - 680 - 690 - 700 - 710 - 720 - 730 - 740 - 750 - 760 - 770 - 780
790 - 800 - 810 - 820 - 830 - 840 - 850 - 860 - 870 - 880 - 890 - 900 - 910
920 - 930 - 940 - 950 - 960 - 970 - 980 - 990 - 1000

Notes: ____

____

____

____

Reg Reps: 10 - 20 - 30 - 40 - 50 - 60 - 70 - 80 - 90 - 100 - 110 - 120 - 130
140 - 150 - 160 - 170 - 180 - 190 - 200 - 210 - 220 - 230 - 240 - 250 - 260
270 - 280 - 290 - 300 - 310 - 320 - 330 - 340 - 350 - 360 - 370 - 380 - 390
400 - 410 - 420 - 430 - 440 - 450 - 460 - 470 - 480 - 490 - 500 - 510 - 520
530 - 540 - 550 - 560 - 570 - 580 - 590 - 600 - 610 - 620 - 630 - 640 - 650
660 - 670 - 680 - 690 - 700 - 710 - 720 - 730 - 740 - 750 - 760 - 770 - 780
790 - 800 - 810 - 820 - 830 - 840 - 850 - 860 - 870 - 880 - 890 - 900 - 910
920 - 930 - 940 - 950 - 960 - 970 - 980 - 990 - 1000

Notes: ____

____

____

____

Reg Reps: 10 - 20 - 30 - 40 - 50 - 60 - 70 - 80 - 90 - 100 - 110 - 120 - 130
140 - 150 - 160 - 170 - 180 - 190 - 200 - 210 - 220 - 230 - 240 - 250 - 260
270 - 280 - 290 - 300 - 310 - 320 - 330 - 340 - 350 - 360 - 370 - 380 - 390
400 - 410 - 420 - 430 - 440 - 450 - 460 - 470 - 480 - 490 - 500 - 510 - 520
530 - 540 - 550 - 560 - 570 - 580 - 590 - 600 - 610 - 620 - 630 - 640 - 650
660 - 670 - 680 - 690 - 700 - 710 - 720 - 730 - 740 - 750 - 760 - 770 - 780
790 - 800 - 810 - 820 - 830 - 840 - 850 - 860 - 870 - 880 - 890 - 900 - 910
920 - 930 - 940 - 950 - 960 - 970 - 980 - 990 - 1000

Notes: ____

____

____

____

Reg Reps: 10 - 20 - 30 - 40 - 50 - 60 - 70 - 80 - 90 - 100 - 110 - 120 - 130
140 - 150 - 160 - 170 - 180 - 190 - 200 - 210 - 220 - 230 - 240 - 250 - 260
270 - 280 - 290 - 300 - 310 - 320 - 330 - 340 - 350 - 360 - 370 - 380 - 390
400 - 410 - 420 - 430 - 440 - 450 - 460 - 470 - 480 - 490 - 500 - 510 - 520
530 - 540 - 550 - 560 - 570 - 580 - 590 - 600 - 610 - 620 - 630 - 640 - 650
660 - 670 - 680 - 690 - 700 - 710 - 720 - 730 - 740 - 750 - 760 - 770 - 780
790 - 800 - 810 - 820 - 830 - 840 - 850 - 860 - 870 - 880 - 890 - 900 - 910
920 - 930 - 940 - 950 - 960 - 970 - 980 - 990 - 1000

Notes: ____

____

____

____

Reg Reps: 10 - 20 - 30 - 40 - 50 - 60 - 70 - 80 - 90 - 100 - 110 - 120 - 130
140 - 150 - 160 - 170 - 180 - 190 - 200 - 210 - 220 - 230 - 240 - 250 - 260
270 - 280 - 290 - 300 - 310 - 320 - 330 - 340 - 350 - 360 - 370 - 380 - 390
400 - 410 - 420 - 430 - 440 - 450 - 460 - 470 - 480 - 490 - 500 - 510 - 520
530 - 540 - 550 - 560 - 570 - 580 - 590 - 600 - 610 - 620 - 630 - 640 - 650
660 - 670 - 680 - 690 - 700 - 710 - 720 - 730 - 740 - 750 - 760 - 770 - 780
790 - 800 - 810 - 820 - 830 - 840 - 850 - 860 - 870 - 880 - 890 - 900 - 910
920 - 930 - 940 - 950 - 960 - 970 - 980 - 990 - 1000

Notes: ____

____

____

__________________________________ **End Date:** __________

Technique:________________________ **L - R - Alt. Start Date:** _______
Notes:

Slow Reps: 10 - 20 - 30 - 40 - 50 - 60 - 70 - 80 - 90 - 100 - 110 - 120 - 130
140 - 150 - 160 - 170 - 180 - 190 - 200 - 210 - 220 - 230 - 240 - 250 - 260
270 - 280 - 290 - 300 - 310 - 320 - 330 - 340 - 350 - 360 - 370 - 380 - 390
400 - 410 - 420 - 430 - 440 - 450 - 460 - 470 - 480 - 490 - 500 - 510 - 520
530 - 540 - 550 - 560 - 570 - 580 - 590 - 600 - 610 - 620 - 630 - 640 - 650
660 - 670 - 680 - 690 - 700 - 710 - 720 - 730 - 740 - 750 - 760 - 770 - 780
790 - 800 - 810 - 820 - 830 - 840 - 850 - 860 - 870 - 880 - 890 - 900 - 910
920 - 930 - 940 - 950 - 960 - 970 - 980 - 990 - 1000

Reg Reps: 10 - 20 - 30 - 40 - 50 - 60 - 70 - 80 - 90 - 100 - 110 - 120 - 130
140 - 150 - 160 - 170 - 180 - 190 - 200 - 210 - 220 - 230 - 240 - 250 - 260
270 - 280 - 290 - 300 - 310 - 320 - 330 - 340 - 350 - 360 - 370 - 380 - 390
400 - 410 - 420 - 430 - 440 - 450 - 460 - 470 - 480 - 490 - 500 - 510 - 520
530 - 540 - 550 - 560 - 570 - 580 - 590 - 600 - 610 - 620 - 630 - 640 - 650
660 - 670 - 680 - 690 - 700 - 710 - 720 - 730 - 740 - 750 - 760 - 770 - 780
790 - 800 - 810 - 820 - 830 - 840 - 850 - 860 - 870 - 880 - 890 - 900 - 910
920 - 930 - 940 - 950 - 960 - 970 - 980 - 990 - 1000
Notes:

Reg Reps: 10 - 20 - 30 - 40 - 50 - 60 - 70 - 80 - 90 - 100 - 110 - 120 - 130
140 - 150 - 160 - 170 - 180 - 190 - 200 - 210 - 220 - 230 - 240 - 250 - 260
270 - 280 - 290 - 300 - 310 - 320 - 330 - 340 - 350 - 360 - 370 - 380 - 390
400 - 410 - 420 - 430 - 440 - 450 - 460 - 470 - 480 - 490 - 500 - 510 - 520
530 - 540 - 550 - 560 - 570 - 580 - 590 - 600 - 610 - 620 - 630 - 640 - 650
660 - 670 - 680 - 690 - 700 - 710 - 720 - 730 - 740 - 750 - 760 - 770 - 780
790 - 800 - 810 - 820 - 830 - 840 - 850 - 860 - 870 - 880 - 890 - 900 - 910
920 - 930 - 940 - 950 - 960 - 970 - 980 - 990 - 1000
Notes:

Reg Reps: 10 - 20 - 30 - 40 - 50 - 60 - 70 - 80 - 90 - 100 - 110 - 120 - 130
140 - 150 - 160 - 170 - 180 - 190 - 200 - 210 - 220 - 230 - 240 - 250 - 260
270 - 280 - 290 - 300 - 310 - 320 - 330 - 340 - 350 - 360 - 370 - 380 - 390
400 - 410 - 420 - 430 - 440 - 450 - 460 - 470 - 480 - 490 - 500 - 510 - 520
530 - 540 - 550 - 560 - 570 - 580 - 590 - 600 - 610 - 620 - 630 - 640 - 650
660 - 670 - 680 - 690 - 700 - 710 - 720 - 730 - 740 - 750 - 760 - 770 - 780
790 - 800 - 810 - 820 - 830 - 840 - 850 - 860 - 870 - 880 - 890 - 900 - 910
920 - 930 - 940 - 950 - 960 - 970 - 980 - 990 - 1000

Notes: ____________________

Reg Reps: 10 - 20 - 30 - 40 - 50 - 60 - 70 - 80 - 90 - 100 - 110 - 120 - 130
140 - 150 - 160 - 170 - 180 - 190 - 200 - 210 - 220 - 230 - 240 - 250 - 260
270 - 280 - 290 - 300 - 310 - 320 - 330 - 340 - 350 - 360 - 370 - 380 - 390
400 - 410 - 420 - 430 - 440 - 450 - 460 - 470 - 480 - 490 - 500 - 510 - 520
530 - 540 - 550 - 560 - 570 - 580 - 590 - 600 - 610 - 620 - 630 - 640 - 650
660 - 670 - 680 - 690 - 700 - 710 - 720 - 730 - 740 - 750 - 760 - 770 - 780
790 - 800 - 810 - 820 - 830 - 840 - 850 - 860 - 870 - 880 - 890 - 900 - 910
920 - 930 - 940 - 950 - 960 - 970 - 980 - 990 - 1000

Notes: ____________________

Reg Reps: 10 - 20 - 30 - 40 - 50 - 60 - 70 - 80 - 90 - 100 - 110 - 120 - 130
140 - 150 - 160 - 170 - 180 - 190 - 200 - 210 - 220 - 230 - 240 - 250 - 260
270 - 280 - 290 - 300 - 310 - 320 - 330 - 340 - 350 - 360 - 370 - 380 - 390
400 - 410 - 420 - 430 - 440 - 450 - 460 - 470 - 480 - 490 - 500 - 510 - 520
530 - 540 - 550 - 560 - 570 - 580 - 590 - 600 - 610 - 620 - 630 - 640 - 650
660 - 670 - 680 - 690 - 700 - 710 - 720 - 730 - 740 - 750 - 760 - 770 - 780
790 - 800 - 810 - 820 - 830 - 840 - 850 - 860 - 870 - 880 - 890 - 900 - 910
920 - 930 - 940 - 950 - 960 - 970 - 980 - 990 - 1000

Notes: ____________________

End Date: __________

Technique: ______________________ **L - R - Alt. Start Date:** _______
Notes: ____

____

____

____

Slow Reps: 10 - 20 - 30 - 40 - 50 - 60 - 70 - 80 - 90 - 100 - 110 - 120 - 130
140 - 150 - 160 - 170 - 180 - 190 - 200 - 210 - 220 - 230 - 240 - 250 - 260
270 - 280 - 290 - 300 - 310 - 320 - 330 - 340 - 350 - 360 - 370 - 380 - 390
400 - 410 - 420 - 430 - 440 - 450 - 460 - 470 - 480 - 490 - 500 - 510 - 520
530 - 540 - 550 - 560 - 570 - 580 - 590 - 600 - 610 - 620 - 630 - 640 - 650
660 - 670 - 680 - 690 - 700 - 710 - 720 - 730 - 740 - 750 - 760 - 770 - 780
790 - 800 - 810 - 820 - 830 - 840 - 850 - 860 - 870 - 880 - 890 - 900 - 910
920 - 930 - 940 - 950 - 960 - 970 - 980 - 990 - 1000

Reg Reps: 10 - 20 - 30 - 40 - 50 - 60 - 70 - 80 - 90 - 100 - 110 - 120 - 130
140 - 150 - 160 - 170 - 180 - 190 - 200 - 210 - 220 - 230 - 240 - 250 - 260
270 - 280 - 290 - 300 - 310 - 320 - 330 - 340 - 350 - 360 - 370 - 380 - 390
400 - 410 - 420 - 430 - 440 - 450 - 460 - 470 - 480 - 490 - 500 - 510 - 520
530 - 540 - 550 - 560 - 570 - 580 - 590 - 600 - 610 - 620 - 630 - 640 - 650
660 - 670 - 680 - 690 - 700 - 710 - 720 - 730 - 740 - 750 - 760 - 770 - 780
790 - 800 - 810 - 820 - 830 - 840 - 850 - 860 - 870 - 880 - 890 - 900 - 910
920 - 930 - 940 - 950 - 960 - 970 - 980 - 990 - 1000
Notes: ____

____

____

____

Reg Reps: 10 - 20 - 30 - 40 - 50 - 60 - 70 - 80 - 90 - 100 - 110 - 120 - 130
140 - 150 - 160 - 170 - 180 - 190 - 200 - 210 - 220 - 230 - 240 - 250 - 260
270 - 280 - 290 - 300 - 310 - 320 - 330 - 340 - 350 - 360 - 370 - 380 - 390
400 - 410 - 420 - 430 - 440 - 450 - 460 - 470 - 480 - 490 - 500 - 510 - 520
530 - 540 - 550 - 560 - 570 - 580 - 590 - 600 - 610 - 620 - 630 - 640 - 650
660 - 670 - 680 - 690 - 700 - 710 - 720 - 730 - 740 - 750 - 760 - 770 - 780
790 - 800 - 810 - 820 - 830 - 840 - 850 - 860 - 870 - 880 - 890 - 900 - 910
920 - 930 - 940 - 950 - 960 - 970 - 980 - 990 - 1000
Notes: ____

____

____

____

Reg Reps: 10 - 20 - 30 - 40 - 50 - 60 - 70 - 80 - 90 - 100 - 110 - 120 - 130
140 - 150 - 160 - 170 - 180 - 190 - 200 - 210 - 220 - 230 - 240 - 250 - 260
270 - 280 - 290 - 300 - 310 - 320 - 330 - 340 - 350 - 360 - 370 - 380 - 390
400 - 410 - 420 - 430 - 440 - 450 - 460 - 470 - 480 - 490 - 500 - 510 - 520
530 - 540 - 550 - 560 - 570 - 580 - 590 - 600 - 610 - 620 - 630 - 640 - 650
660 - 670 - 680 - 690 - 700 - 710 - 720 - 730 - 740 - 750 - 760 - 770 - 780
790 - 800 - 810 - 820 - 830 - 840 - 850 - 860 - 870 - 880 - 890 - 900 - 910
920 - 930 - 940 - 950 - 960 - 970 - 980 - 990 - 1000

Notes: ______________________________

____

____

____

Reg Reps: 10 - 20 - 30 - 40 - 50 - 60 - 70 - 80 - 90 - 100 - 110 - 120 - 130
140 - 150 - 160 - 170 - 180 - 190 - 200 - 210 - 220 - 230 - 240 - 250 - 260
270 - 280 - 290 - 300 - 310 - 320 - 330 - 340 - 350 - 360 - 370 - 380 - 390
400 - 410 - 420 - 430 - 440 - 450 - 460 - 470 - 480 - 490 - 500 - 510 - 520
530 - 540 - 550 - 560 - 570 - 580 - 590 - 600 - 610 - 620 - 630 - 640 - 650
660 - 670 - 680 - 690 - 700 - 710 - 720 - 730 - 740 - 750 - 760 - 770 - 780
790 - 800 - 810 - 820 - 830 - 840 - 850 - 860 - 870 - 880 - 890 - 900 - 910
920 - 930 - 940 - 950 - 960 - 970 - 980 - 990 - 1000

Notes: ______________________________

____

____

____

Reg Reps: 10 - 20 - 30 - 40 - 50 - 60 - 70 - 80 - 90 - 100 - 110 - 120 - 130
140 - 150 - 160 - 170 - 180 - 190 - 200 - 210 - 220 - 230 - 240 - 250 - 260
270 - 280 - 290 - 300 - 310 - 320 - 330 - 340 - 350 - 360 - 370 - 380 - 390
400 - 410 - 420 - 430 - 440 - 450 - 460 - 470 - 480 - 490 - 500 - 510 - 520
530 - 540 - 550 - 560 - 570 - 580 - 590 - 600 - 610 - 620 - 630 - 640 - 650
660 - 670 - 680 - 690 - 700 - 710 - 720 - 730 - 740 - 750 - 760 - 770 - 780
790 - 800 - 810 - 820 - 830 - 840 - 850 - 860 - 870 - 880 - 890 - 900 - 910
920 - 930 - 940 - 950 - 960 - 970 - 980 - 990 - 1000

Notes: ______________________________

____

____

______________________________**End Date:** ______________

Technique:________________________ **L - R - Alt. Start Date:** _______

Notes:____

Slow Reps: 10 - 20 - 30 - 40 - 50 - 60 - 70 - 80 - 90 - 100 - 110 - 120 - 130
140 - 150 - 160 - 170 - 180 - 190 - 200 - 210 - 220 - 230 - 240 - 250 - 260
270 - 280 - 290 - 300 - 310 - 320 - 330 - 340 - 350 - 360 - 370 - 380 - 390
400 - 410 - 420 - 430 - 440 - 450 - 460 - 470 - 480 - 490 - 500 - 510 - 520
530 - 540 - 550 - 560 - 570 - 580 - 590 - 600 - 610 - 620 - 630 - 640 - 650
660 - 670 - 680 - 690 - 700 - 710 - 720 - 730 - 740 - 750 - 760 - 770 - 780
790 - 800 - 810 - 820 - 830 - 840 - 850 - 860 - 870 - 880 - 890 - 900 - 910
920 - 930 - 940 - 950 - 960 - 970 - 980 - 990 - 1000

Reg Reps: 10 - 20 - 30 - 40 - 50 - 60 - 70 - 80 - 90 - 100 - 110 - 120 - 130
140 - 150 - 160 - 170 - 180 - 190 - 200 - 210 - 220 - 230 - 240 - 250 - 260
270 - 280 - 290 - 300 - 310 - 320 - 330 - 340 - 350 - 360 - 370 - 380 - 390
400 - 410 - 420 - 430 - 440 - 450 - 460 - 470 - 480 - 490 - 500 - 510 - 520
530 - 540 - 550 - 560 - 570 - 580 - 590 - 600 - 610 - 620 - 630 - 640 - 650
660 - 670 - 680 - 690 - 700 - 710 - 720 - 730 - 740 - 750 - 760 - 770 - 780
790 - 800 - 810 - 820 - 830 - 840 - 850 - 860 - 870 - 880 - 890 - 900 - 910
920 - 930 - 940 - 950 - 960 - 970 - 980 - 990 - 1000

Notes: ____

Reg Reps: 10 - 20 - 30 - 40 - 50 - 60 - 70 - 80 - 90 - 100 - 110 - 120 - 130
140 - 150 - 160 - 170 - 180 - 190 - 200 - 210 - 220 - 230 - 240 - 250 - 260
270 - 280 - 290 - 300 - 310 - 320 - 330 - 340 - 350 - 360 - 370 - 380 - 390
400 - 410 - 420 - 430 - 440 - 450 - 460 - 470 - 480 - 490 - 500 - 510 - 520
530 - 540 - 550 - 560 - 570 - 580 - 590 - 600 - 610 - 620 - 630 - 640 - 650
660 - 670 - 680 - 690 - 700 - 710 - 720 - 730 - 740 - 750 - 760 - 770 - 780
790 - 800 - 810 - 820 - 830 - 840 - 850 - 860 - 870 - 880 - 890 - 900 - 910
920 - 930 - 940 - 950 - 960 - 970 - 980 - 990 - 1000

Notes: ____

Reg Reps: 10 - 20 - 30 - 40 - 50 - 60 - 70 - 80 - 90 - 100 - 110 - 120 - 130 140 - 150 - 160 - 170 - 180 - 190 - 200 - 210 - 220 - 230 - 240 - 250 - 260 270 - 280 - 290 - 300 - 310 - 320 - 330 - 340 - 350 - 360 - 370 - 380 - 390 400 - 410 - 420 - 430 - 440 - 450 - 460 - 470 - 480 - 490 - 500 - 510 - 520 530 - 540 - 550 - 560 - 570 - 580 - 590 - 600 - 610 - 620 - 630 - 640 - 650 660 - 670 - 680 - 690 - 700 - 710 - 720 - 730 - 740 - 750 - 760 - 770 - 780 790 - 800 - 810 - 820 - 830 - 840 - 850 - 860 - 870 - 880 - 890 - 900 - 910 920 - 930 - 940 - 950 - 960 - 970 - 980 - 990 - 1000

Notes: ____

____

____

____

Reg Reps: 10 - 20 - 30 - 40 - 50 - 60 - 70 - 80 - 90 - 100 - 110 - 120 - 130 140 - 150 - 160 - 170 - 180 - 190 - 200 - 210 - 220 - 230 - 240 - 250 - 260 270 - 280 - 290 - 300 - 310 - 320 - 330 - 340 - 350 - 360 - 370 - 380 - 390 400 - 410 - 420 - 430 - 440 - 450 - 460 - 470 - 480 - 490 - 500 - 510 - 520 530 - 540 - 550 - 560 - 570 - 580 - 590 - 600 - 610 - 620 - 630 - 640 - 650 660 - 670 - 680 - 690 - 700 - 710 - 720 - 730 - 740 - 750 - 760 - 770 - 780 790 - 800 - 810 - 820 - 830 - 840 - 850 - 860 - 870 - 880 - 890 - 900 - 910 920 - 930 - 940 - 950 - 960 - 970 - 980 - 990 - 1000

Notes: ____

____

____

____

Reg Reps: 10 - 20 - 30 - 40 - 50 - 60 - 70 - 80 - 90 - 100 - 110 - 120 - 130 140 - 150 - 160 - 170 - 180 - 190 - 200 - 210 - 220 - 230 - 240 - 250 - 260 270 - 280 - 290 - 300 - 310 - 320 - 330 - 340 - 350 - 360 - 370 - 380 - 390 400 - 410 - 420 - 430 - 440 - 450 - 460 - 470 - 480 - 490 - 500 - 510 - 520 530 - 540 - 550 - 560 - 570 - 580 - 590 - 600 - 610 - 620 - 630 - 640 - 650 660 - 670 - 680 - 690 - 700 - 710 - 720 - 730 - 740 - 750 - 760 - 770 - 780 790 - 800 - 810 - 820 - 830 - 840 - 850 - 860 - 870 - 880 - 890 - 900 - 910 920 - 930 - 940 - 950 - 960 - 970 - 980 - 990 - 1000

Notes: ____

____

____

____ **End Date:**____________

Technique:________________________ **L - R - Alt. Start Date:** _______

Notes:____

Slow Reps: 10 - 20 - 30 - 40 - 50 - 60 - 70 - 80 - 90 - 100 - 110 - 120 - 130
140 - 150 - 160 - 170 - 180 - 190 - 200 - 210 - 220 - 230 - 240 - 250 - 260
270 - 280 - 290 - 300 - 310 - 320 - 330 - 340 - 350 - 360 - 370 - 380 - 390
400 - 410 - 420 - 430 - 440 - 450 - 460 - 470 - 480 - 490 - 500 - 510 - 520
530 - 540 - 550 - 560 - 570 - 580 - 590 - 600 - 610 - 620 - 630 - 640 - 650
660 - 670 - 680 - 690 - 700 - 710 - 720 - 730 - 740 - 750 - 760 - 770 - 780
790 - 800 - 810 - 820 - 830 - 840 - 850 - 860 - 870 - 880 - 890 - 900 - 910
920 - 930 - 940 - 950 - 960 - 970 - 980 - 990 - 1000

Reg Reps: 10 - 20 - 30 - 40 - 50 - 60 - 70 - 80 - 90 - 100 - 110 - 120 - 130
140 - 150 - 160 - 170 - 180 - 190 - 200 - 210 - 220 - 230 - 240 - 250 - 260
270 - 280 - 290 - 300 - 310 - 320 - 330 - 340 - 350 - 360 - 370 - 380 - 390
400 - 410 - 420 - 430 - 440 - 450 - 460 - 470 - 480 - 490 - 500 - 510 - 520
530 - 540 - 550 - 560 - 570 - 580 - 590 - 600 - 610 - 620 - 630 - 640 - 650
660 - 670 - 680 - 690 - 700 - 710 - 720 - 730 - 740 - 750 - 760 - 770 - 780
790 - 800 - 810 - 820 - 830 - 840 - 850 - 860 - 870 - 880 - 890 - 900 - 910
920 - 930 - 940 - 950 - 960 - 970 - 980 - 990 - 1000

Notes: ____

Reg Reps: 10 - 20 - 30 - 40 - 50 - 60 - 70 - 80 - 90 - 100 - 110 - 120 - 130
140 - 150 - 160 - 170 - 180 - 190 - 200 - 210 - 220 - 230 - 240 - 250 - 260
270 - 280 - 290 - 300 - 310 - 320 - 330 - 340 - 350 - 360 - 370 - 380 - 390
400 - 410 - 420 - 430 - 440 - 450 - 460 - 470 - 480 - 490 - 500 - 510 - 520
530 - 540 - 550 - 560 - 570 - 580 - 590 - 600 - 610 - 620 - 630 - 640 - 650
660 - 670 - 680 - 690 - 700 - 710 - 720 - 730 - 740 - 750 - 760 - 770 - 780
790 - 800 - 810 - 820 - 830 - 840 - 850 - 860 - 870 - 880 - 890 - 900 - 910
920 - 930 - 940 - 950 - 960 - 970 - 980 - 990 - 1000

Notes: ____

Reg Reps: 10 - 20 - 30 - 40 - 50 - 60 - 70 - 80 - 90 - 100 - 110 - 120 - 130
140 - 150 - 160 - 170 - 180 - 190 - 200 - 210 - 220 - 230 - 240 - 250 - 260
270 - 280 - 290 - 300 - 310 - 320 - 330 - 340 - 350 - 360 - 370 - 380 - 390
400 - 410 - 420 - 430 - 440 - 450 - 460 - 470 - 480 - 490 - 500 - 510 - 520
530 - 540 - 550 - 560 - 570 - 580 - 590 - 600 - 610 - 620 - 630 - 640 - 650
660 - 670 - 680 - 690 - 700 - 710 - 720 - 730 - 740 - 750 - 760 - 770 - 780
790 - 800 - 810 - 820 - 830 - 840 - 850 - 860 - 870 - 880 - 890 - 900 - 910
920 - 930 - 940 - 950 - 960 - 970 - 980 - 990 - 1000

Notes: ______________________________

____

____

____

Reg Reps: 10 - 20 - 30 - 40 - 50 - 60 - 70 - 80 - 90 - 100 - 110 - 120 - 130
140 - 150 - 160 - 170 - 180 - 190 - 200 - 210 - 220 - 230 - 240 - 250 - 260
270 - 280 - 290 - 300 - 310 - 320 - 330 - 340 - 350 - 360 - 370 - 380 - 390
400 - 410 - 420 - 430 - 440 - 450 - 460 - 470 - 480 - 490 - 500 - 510 - 520
530 - 540 - 550 - 560 - 570 - 580 - 590 - 600 - 610 - 620 - 630 - 640 - 650
660 - 670 - 680 - 690 - 700 - 710 - 720 - 730 - 740 - 750 - 760 - 770 - 780
790 - 800 - 810 - 820 - 830 - 840 - 850 - 860 - 870 - 880 - 890 - 900 - 910
920 - 930 - 940 - 950 - 960 - 970 - 980 - 990 - 1000

Notes: ______________________________

____

____

____

Reg Reps: 10 - 20 - 30 - 40 - 50 - 60 - 70 - 80 - 90 - 100 - 110 - 120 - 130
140 - 150 - 160 - 170 - 180 - 190 - 200 - 210 - 220 - 230 - 240 - 250 - 260
270 - 280 - 290 - 300 - 310 - 320 - 330 - 340 - 350 - 360 - 370 - 380 - 390
400 - 410 - 420 - 430 - 440 - 450 - 460 - 470 - 480 - 490 - 500 - 510 - 520
530 - 540 - 550 - 560 - 570 - 580 - 590 - 600 - 610 - 620 - 630 - 640 - 650
660 - 670 - 680 - 690 - 700 - 710 - 720 - 730 - 740 - 750 - 760 - 770 - 780
790 - 800 - 810 - 820 - 830 - 840 - 850 - 860 - 870 - 880 - 890 - 900 - 910
920 - 930 - 940 - 950 - 960 - 970 - 980 - 990 - 1000

Notes: ______________________________

____

____

______________________________ **End Date:** ____________

Technique: ______________________ **L - R - Alt. Start Date:** _______

Notes: ____

____

____

____

Slow Reps: 10 - 20 - 30 - 40 - 50 - 60 - 70 - 80 - 90 - 100 - 110 - 120 - 130
140 - 150 - 160 - 170 - 180 - 190 - 200 - 210 - 220 - 230 - 240 - 250 - 260
270 - 280 - 290 - 300 - 310 - 320 - 330 - 340 - 350 - 360 - 370 - 380 - 390
400 - 410 - 420 - 430 - 440 - 450 - 460 - 470 - 480 - 490 - 500 - 510 - 520
530 - 540 - 550 - 560 - 570 - 580 - 590 - 600 - 610 - 620 - 630 - 640 - 650
660 - 670 - 680 - 690 - 700 - 710 - 720 - 730 - 740 - 750 - 760 - 770 - 780
790 - 800 - 810 - 820 - 830 - 840 - 850 - 860 - 870 - 880 - 890 - 900 - 910
920 - 930 - 940 - 950 - 960 - 970 - 980 - 990 - 1000

Reg Reps: 10 - 20 - 30 - 40 - 50 - 60 - 70 - 80 - 90 - 100 - 110 - 120 - 130
140 - 150 - 160 - 170 - 180 - 190 - 200 - 210 - 220 - 230 - 240 - 250 - 260
270 - 280 - 290 - 300 - 310 - 320 - 330 - 340 - 350 - 360 - 370 - 380 - 390
400 - 410 - 420 - 430 - 440 - 450 - 460 - 470 - 480 - 490 - 500 - 510 - 520
530 - 540 - 550 - 560 - 570 - 580 - 590 - 600 - 610 - 620 - 630 - 640 - 650
660 - 670 - 680 - 690 - 700 - 710 - 720 - 730 - 740 - 750 - 760 - 770 - 780
790 - 800 - 810 - 820 - 830 - 840 - 850 - 860 - 870 - 880 - 890 - 900 - 910
920 - 930 - 940 - 950 - 960 - 970 - 980 - 990 - 1000

Notes: ____

____

____

____

Reg Reps: 10 - 20 - 30 - 40 - 50 - 60 - 70 - 80 - 90 - 100 - 110 - 120 - 130
140 - 150 - 160 - 170 - 180 - 190 - 200 - 210 - 220 - 230 - 240 - 250 - 260
270 - 280 - 290 - 300 - 310 - 320 - 330 - 340 - 350 - 360 - 370 - 380 - 390
400 - 410 - 420 - 430 - 440 - 450 - 460 - 470 - 480 - 490 - 500 - 510 - 520
530 - 540 - 550 - 560 - 570 - 580 - 590 - 600 - 610 - 620 - 630 - 640 - 650
660 - 670 - 680 - 690 - 700 - 710 - 720 - 730 - 740 - 750 - 760 - 770 - 780
790 - 800 - 810 - 820 - 830 - 840 - 850 - 860 - 870 - 880 - 890 - 900 - 910
920 - 930 - 940 - 950 - 960 - 970 - 980 - 990 - 1000

Notes: ____

____

____

____

Reg Reps: 10 - 20 - 30 - 40 - 50 - 60 - 70 - 80 - 90 - 100 - 110 - 120 - 130
140 - 150 - 160 - 170 - 180 - 190 - 200 - 210 - 220 - 230 - 240 - 250 - 260
270 - 280 - 290 - 300 - 310 - 320 - 330 - 340 - 350 - 360 - 370 - 380 - 390
400 - 410 - 420 - 430 - 440 - 450 - 460 - 470 - 480 - 490 - 500 - 510 - 520
530 - 540 - 550 - 560 - 570 - 580 - 590 - 600 - 610 - 620 - 630 - 640 - 650
660 - 670 - 680 - 690 - 700 - 710 - 720 - 730 - 740 - 750 - 760 - 770 - 780
790 - 800 - 810 - 820 - 830 - 840 - 850 - 860 - 870 - 880 - 890 - 900 - 910
920 - 930 - 940 - 950 - 960 - 970 - 980 - 990 - 1000

Notes: ____

____

____

____

Reg Reps: 10 - 20 - 30 - 40 - 50 - 60 - 70 - 80 - 90 - 100 - 110 - 120 - 130
140 - 150 - 160 - 170 - 180 - 190 - 200 - 210 - 220 - 230 - 240 - 250 - 260
270 - 280 - 290 - 300 - 310 - 320 - 330 - 340 - 350 - 360 - 370 - 380 - 390
400 - 410 - 420 - 430 - 440 - 450 - 460 - 470 - 480 - 490 - 500 - 510 - 520
530 - 540 - 550 - 560 - 570 - 580 - 590 - 600 - 610 - 620 - 630 - 640 - 650
660 - 670 - 680 - 690 - 700 - 710 - 720 - 730 - 740 - 750 - 760 - 770 - 780
790 - 800 - 810 - 820 - 830 - 840 - 850 - 860 - 870 - 880 - 890 - 900 - 910
920 - 930 - 940 - 950 - 960 - 970 - 980 - 990 - 1000

Notes: ____

____

____

____

Reg Reps: 10 - 20 - 30 - 40 - 50 - 60 - 70 - 80 - 90 - 100 - 110 - 120 - 130
140 - 150 - 160 - 170 - 180 - 190 - 200 - 210 - 220 - 230 - 240 - 250 - 260
270 - 280 - 290 - 300 - 310 - 320 - 330 - 340 - 350 - 360 - 370 - 380 - 390
400 - 410 - 420 - 430 - 440 - 450 - 460 - 470 - 480 - 490 - 500 - 510 - 520
530 - 540 - 550 - 560 - 570 - 580 - 590 - 600 - 610 - 620 - 630 - 640 - 650
660 - 670 - 680 - 690 - 700 - 710 - 720 - 730 - 740 - 750 - 760 - 770 - 780
790 - 800 - 810 - 820 - 830 - 840 - 850 - 860 - 870 - 880 - 890 - 900 - 910
920 - 930 - 940 - 950 - 960 - 970 - 980 - 990 - 1000

Notes: ____

____

____

______________________________ **End Date:** ____________

Technique:________________________ **L - R - Alt. Start Date:** _______
Notes:____

Slow Reps: 10 - 20 - 30 - 40 - 50 - 60 - 70 - 80 - 90 - 100 - 110 - 120 - 130
140 - 150 - 160 - 170 - 180 - 190 - 200 - 210 - 220 - 230 - 240 - 250 - 260
270 - 280 - 290 - 300 - 310 - 320 - 330 - 340 - 350 - 360 - 370 - 380 - 390
400 - 410 - 420 - 430 - 440 - 450 - 460 - 470 - 480 - 490 - 500 - 510 - 520
530 - 540 - 550 - 560 - 570 - 580 - 590 - 600 - 610 - 620 - 630 - 640 - 650
660 - 670 - 680 - 690 - 700 - 710 - 720 - 730 - 740 - 750 - 760 - 770 - 780
790 - 800 - 810 - 820 - 830 - 840 - 850 - 860 - 870 - 880 - 890 - 900 - 910
920 - 930 - 940 - 950 - 960 - 970 - 980 - 990 - 1000

Reg Reps: 10 - 20 - 30 - 40 - 50 - 60 - 70 - 80 - 90 - 100 - 110 - 120 - 130
140 - 150 - 160 - 170 - 180 - 190 - 200 - 210 - 220 - 230 - 240 - 250 - 260
270 - 280 - 290 - 300 - 310 - 320 - 330 - 340 - 350 - 360 - 370 - 380 - 390
400 - 410 - 420 - 430 - 440 - 450 - 460 - 470 - 480 - 490 - 500 - 510 - 520
530 - 540 - 550 - 560 - 570 - 580 - 590 - 600 - 610 - 620 - 630 - 640 - 650
660 - 670 - 680 - 690 - 700 - 710 - 720 - 730 - 740 - 750 - 760 - 770 - 780
790 - 800 - 810 - 820 - 830 - 840 - 850 - 860 - 870 - 880 - 890 - 900 - 910
920 - 930 - 940 - 950 - 960 - 970 - 980 - 990 - 1000
Notes: ____

Reg Reps: 10 - 20 - 30 - 40 - 50 - 60 - 70 - 80 - 90 - 100 - 110 - 120 - 130
140 - 150 - 160 - 170 - 180 - 190 - 200 - 210 - 220 - 230 - 240 - 250 - 260
270 - 280 - 290 - 300 - 310 - 320 - 330 - 340 - 350 - 360 - 370 - 380 - 390
400 - 410 - 420 - 430 - 440 - 450 - 460 - 470 - 480 - 490 - 500 - 510 - 520
530 - 540 - 550 - 560 - 570 - 580 - 590 - 600 - 610 - 620 - 630 - 640 - 650
660 - 670 - 680 - 690 - 700 - 710 - 720 - 730 - 740 - 750 - 760 - 770 - 780
790 - 800 - 810 - 820 - 830 - 840 - 850 - 860 - 870 - 880 - 890 - 900 - 910
920 - 930 - 940 - 950 - 960 - 970 - 980 - 990 - 1000
Notes: ____

Reg Reps: 10 - 20 - 30 - 40 - 50 - 60 - 70 - 80 - 90 - 100 - 110 - 120 - 130
140 - 150 - 160 - 170 - 180 - 190 - 200 - 210 - 220 - 230 - 240 - 250 - 260
270 - 280 - 290 - 300 - 310 - 320 - 330 - 340 - 350 - 360 - 370 - 380 - 390
400 - 410 - 420 - 430 - 440 - 450 - 460 - 470 - 480 - 490 - 500 - 510 - 520
530 - 540 - 550 - 560 - 570 - 580 - 590 - 600 - 610 - 620 - 630 - 640 - 650
660 - 670 - 680 - 690 - 700 - 710 - 720 - 730 - 740 - 750 - 760 - 770 - 780
790 - 800 - 810 - 820 - 830 - 840 - 850 - 860 - 870 - 880 - 890 - 900 - 910
920 - 930 - 940 - 950 - 960 - 970 - 980 - 990 - 1000

Notes: ______________________________

____

____

____

Reg Reps: 10 - 20 - 30 - 40 - 50 - 60 - 70 - 80 - 90 - 100 - 110 - 120 - 130
140 - 150 - 160 - 170 - 180 - 190 - 200 - 210 - 220 - 230 - 240 - 250 - 260
270 - 280 - 290 - 300 - 310 - 320 - 330 - 340 - 350 - 360 - 370 - 380 - 390
400 - 410 - 420 - 430 - 440 - 450 - 460 - 470 - 480 - 490 - 500 - 510 - 520
530 - 540 - 550 - 560 - 570 - 580 - 590 - 600 - 610 - 620 - 630 - 640 - 650
660 - 670 - 680 - 690 - 700 - 710 - 720 - 730 - 740 - 750 - 760 - 770 - 780
790 - 800 - 810 - 820 - 830 - 840 - 850 - 860 - 870 - 880 - 890 - 900 - 910
920 - 930 - 940 - 950 - 960 - 970 - 980 - 990 - 1000

Notes: ______________________________

____

____

____

Reg Reps: 10 - 20 - 30 - 40 - 50 - 60 - 70 - 80 - 90 - 100 - 110 - 120 - 130
140 - 150 - 160 - 170 - 180 - 190 - 200 - 210 - 220 - 230 - 240 - 250 - 260
270 - 280 - 290 - 300 - 310 - 320 - 330 - 340 - 350 - 360 - 370 - 380 - 390
400 - 410 - 420 - 430 - 440 - 450 - 460 - 470 - 480 - 490 - 500 - 510 - 520
530 - 540 - 550 - 560 - 570 - 580 - 590 - 600 - 610 - 620 - 630 - 640 - 650
660 - 670 - 680 - 690 - 700 - 710 - 720 - 730 - 740 - 750 - 760 - 770 - 780
790 - 800 - 810 - 820 - 830 - 840 - 850 - 860 - 870 - 880 - 890 - 900 - 910
920 - 930 - 940 - 950 - 960 - 970 - 980 - 990 - 1000

Notes: ______________________________

____

____

______________________________ **End Date:** ______________

Technique:______________________ **L - R - Alt. Start Date:** _______
Notes:____

Slow Reps: 10 - 20 - 30 - 40 - 50 - 60 - 70 - 80 - 90 - 100 - 110 - 120 - 130
140 - 150 - 160 - 170 - 180 - 190 - 200 - 210 - 220 - 230 - 240 - 250 - 260
270 - 280 - 290 - 300 - 310 - 320 - 330 - 340 - 350 - 360 - 370 - 380 - 390
400 - 410 - 420 - 430 - 440 - 450 - 460 - 470 - 480 - 490 - 500 - 510 - 520
530 - 540 - 550 - 560 - 570 - 580 - 590 - 600 - 610 - 620 - 630 - 640 - 650
660 - 670 - 680 - 690 - 700 - 710 - 720 - 730 - 740 - 750 - 760 - 770 - 780
790 - 800 - 810 - 820 - 830 - 840 - 850 - 860 - 870 - 880 - 890 - 900 - 910
920 - 930 - 940 - 950 - 960 - 970 - 980 - 990 - 1000

Reg Reps: 10 - 20 - 30 - 40 - 50 - 60 - 70 - 80 - 90 - 100 - 110 - 120 - 130
140 - 150 - 160 - 170 - 180 - 190 - 200 - 210 - 220 - 230 - 240 - 250 - 260
270 - 280 - 290 - 300 - 310 - 320 - 330 - 340 - 350 - 360 - 370 - 380 - 390
400 - 410 - 420 - 430 - 440 - 450 - 460 - 470 - 480 - 490 - 500 - 510 - 520
530 - 540 - 550 - 560 - 570 - 580 - 590 - 600 - 610 - 620 - 630 - 640 - 650
660 - 670 - 680 - 690 - 700 - 710 - 720 - 730 - 740 - 750 - 760 - 770 - 780
790 - 800 - 810 - 820 - 830 - 840 - 850 - 860 - 870 - 880 - 890 - 900 - 910
920 - 930 - 940 - 950 - 960 - 970 - 980 - 990 - 1000
Notes: ____

Reg Reps: 10 - 20 - 30 - 40 - 50 - 60 - 70 - 80 - 90 - 100 - 110 - 120 - 130
140 - 150 - 160 - 170 - 180 - 190 - 200 - 210 - 220 - 230 - 240 - 250 - 260
270 - 280 - 290 - 300 - 310 - 320 - 330 - 340 - 350 - 360 - 370 - 380 - 390
400 - 410 - 420 - 430 - 440 - 450 - 460 - 470 - 480 - 490 - 500 - 510 - 520
530 - 540 - 550 - 560 - 570 - 580 - 590 - 600 - 610 - 620 - 630 - 640 - 650
660 - 670 - 680 - 690 - 700 - 710 - 720 - 730 - 740 - 750 - 760 - 770 - 780
790 - 800 - 810 - 820 - 830 - 840 - 850 - 860 - 870 - 880 - 890 - 900 - 910
920 - 930 - 940 - 950 - 960 - 970 - 980 - 990 - 1000
Notes: ____

Reg Reps: 10 - 20 - 30 - 40 - 50 - 60 - 70 - 80 - 90 - 100 - 110 - 120 - 130
140 - 150 - 160 - 170 - 180 - 190 - 200 - 210 - 220 - 230 - 240 - 250 - 260
270 - 280 - 290 - 300 - 310 - 320 - 330 - 340 - 350 - 360 - 370 - 380 - 390
400 - 410 - 420 - 430 - 440 - 450 - 460 - 470 - 480 - 490 - 500 - 510 - 520
530 - 540 - 550 - 560 - 570 - 580 - 590 - 600 - 610 - 620 - 630 - 640 - 650
660 - 670 - 680 - 690 - 700 - 710 - 720 - 730 - 740 - 750 - 760 - 770 - 780
790 - 800 - 810 - 820 - 830 - 840 - 850 - 860 - 870 - 880 - 890 - 900 - 910
920 - 930 - 940 - 950 - 960 - 970 - 980 - 990 - 1000

Notes: ____

____

____

____

Reg Reps: 10 - 20 - 30 - 40 - 50 - 60 - 70 - 80 - 90 - 100 - 110 - 120 - 130
140 - 150 - 160 - 170 - 180 - 190 - 200 - 210 - 220 - 230 - 240 - 250 - 260
270 - 280 - 290 - 300 - 310 - 320 - 330 - 340 - 350 - 360 - 370 - 380 - 390
400 - 410 - 420 - 430 - 440 - 450 - 460 - 470 - 480 - 490 - 500 - 510 - 520
530 - 540 - 550 - 560 - 570 - 580 - 590 - 600 - 610 - 620 - 630 - 640 - 650
660 - 670 - 680 - 690 - 700 - 710 - 720 - 730 - 740 - 750 - 760 - 770 - 780
790 - 800 - 810 - 820 - 830 - 840 - 850 - 860 - 870 - 880 - 890 - 900 - 910
920 - 930 - 940 - 950 - 960 - 970 - 980 - 990 - 1000

Notes: ____

____

____

____

Reg Reps: 10 - 20 - 30 - 40 - 50 - 60 - 70 - 80 - 90 - 100 - 110 - 120 - 130
140 - 150 - 160 - 170 - 180 - 190 - 200 - 210 - 220 - 230 - 240 - 250 - 260
270 - 280 - 290 - 300 - 310 - 320 - 330 - 340 - 350 - 360 - 370 - 380 - 390
400 - 410 - 420 - 430 - 440 - 450 - 460 - 470 - 480 - 490 - 500 - 510 - 520
530 - 540 - 550 - 560 - 570 - 580 - 590 - 600 - 610 - 620 - 630 - 640 - 650
660 - 670 - 680 - 690 - 700 - 710 - 720 - 730 - 740 - 750 - 760 - 770 - 780
790 - 800 - 810 - 820 - 830 - 840 - 850 - 860 - 870 - 880 - 890 - 900 - 910
920 - 930 - 940 - 950 - 960 - 970 - 980 - 990 - 1000

Notes: ____

____

____

______________________________________ **End Date:** __________

Technique: ______________________ **L - R - Alt. Start Date:** _______

Notes: ____

____

____

____

Slow Reps: 10 - 20 - 30 - 40 - 50 - 60 - 70 - 80 - 90 - 100 - 110 - 120 - 130
140 - 150 - 160 - 170 - 180 - 190 - 200 - 210 - 220 - 230 - 240 - 250 - 260
270 - 280 - 290 - 300 - 310 - 320 - 330 - 340 - 350 - 360 - 370 - 380 - 390
400 - 410 - 420 - 430 - 440 - 450 - 460 - 470 - 480 - 490 - 500 - 510 - 520
530 - 540 - 550 - 560 - 570 - 580 - 590 - 600 - 610 - 620 - 630 - 640 - 650
660 - 670 - 680 - 690 - 700 - 710 - 720 - 730 - 740 - 750 - 760 - 770 - 780
790 - 800 - 810 - 820 - 830 - 840 - 850 - 860 - 870 - 880 - 890 - 900 - 910
920 - 930 - 940 - 950 - 960 - 970 - 980 - 990 - 1000

Reg Reps: 10 - 20 - 30 - 40 - 50 - 60 - 70 - 80 - 90 - 100 - 110 - 120 - 130
140 - 150 - 160 - 170 - 180 - 190 - 200 - 210 - 220 - 230 - 240 - 250 - 260
270 - 280 - 290 - 300 - 310 - 320 - 330 - 340 - 350 - 360 - 370 - 380 - 390
400 - 410 - 420 - 430 - 440 - 450 - 460 - 470 - 480 - 490 - 500 - 510 - 520
530 - 540 - 550 - 560 - 570 - 580 - 590 - 600 - 610 - 620 - 630 - 640 - 650
660 - 670 - 680 - 690 - 700 - 710 - 720 - 730 - 740 - 750 - 760 - 770 - 780
790 - 800 - 810 - 820 - 830 - 840 - 850 - 860 - 870 - 880 - 890 - 900 - 910
920 - 930 - 940 - 950 - 960 - 970 - 980 - 990 - 1000

Notes: ____

____

____

____

Reg Reps: 10 - 20 - 30 - 40 - 50 - 60 - 70 - 80 - 90 - 100 - 110 - 120 - 130
140 - 150 - 160 - 170 - 180 - 190 - 200 - 210 - 220 - 230 - 240 - 250 - 260
270 - 280 - 290 - 300 - 310 - 320 - 330 - 340 - 350 - 360 - 370 - 380 - 390
400 - 410 - 420 - 430 - 440 - 450 - 460 - 470 - 480 - 490 - 500 - 510 - 520
530 - 540 - 550 - 560 - 570 - 580 - 590 - 600 - 610 - 620 - 630 - 640 - 650
660 - 670 - 680 - 690 - 700 - 710 - 720 - 730 - 740 - 750 - 760 - 770 - 780
790 - 800 - 810 - 820 - 830 - 840 - 850 - 860 - 870 - 880 - 890 - 900 - 910
920 - 930 - 940 - 950 - 960 - 970 - 980 - 990 - 1000

Notes: ____

____

____

____

Reg Reps: 10 - 20 - 30 - 40 - 50 - 60 - 70 - 80 - 90 - 100 - 110 - 120 - 130
140 - 150 - 160 - 170 - 180 - 190 - 200 - 210 - 220 - 230 - 240 - 250 - 260
270 - 280 - 290 - 300 - 310 - 320 - 330 - 340 - 350 - 360 - 370 - 380 - 390
400 - 410 - 420 - 430 - 440 - 450 - 460 - 470 - 480 - 490 - 500 - 510 - 520
530 - 540 - 550 - 560 - 570 - 580 - 590 - 600 - 610 - 620 - 630 - 640 - 650
660 - 670 - 680 - 690 - 700 - 710 - 720 - 730 - 740 - 750 - 760 - 770 - 780
790 - 800 - 810 - 820 - 830 - 840 - 850 - 860 - 870 - 880 - 890 - 900 - 910
920 - 930 - 940 - 950 - 960 - 970 - 980 - 990 - 1000

Notes: ____

____

____

____

Reg Reps: 10 - 20 - 30 - 40 - 50 - 60 - 70 - 80 - 90 - 100 - 110 - 120 - 130
140 - 150 - 160 - 170 - 180 - 190 - 200 - 210 - 220 - 230 - 240 - 250 - 260
270 - 280 - 290 - 300 - 310 - 320 - 330 - 340 - 350 - 360 - 370 - 380 - 390
400 - 410 - 420 - 430 - 440 - 450 - 460 - 470 - 480 - 490 - 500 - 510 - 520
530 - 540 - 550 - 560 - 570 - 580 - 590 - 600 - 610 - 620 - 630 - 640 - 650
660 - 670 - 680 - 690 - 700 - 710 - 720 - 730 - 740 - 750 - 760 - 770 - 780
790 - 800 - 810 - 820 - 830 - 840 - 850 - 860 - 870 - 880 - 890 - 900 - 910
920 - 930 - 940 - 950 - 960 - 970 - 980 - 990 - 1000

Notes: ____

____

____

____

Reg Reps: 10 - 20 - 30 - 40 - 50 - 60 - 70 - 80 - 90 - 100 - 110 - 120 - 130
140 - 150 - 160 - 170 - 180 - 190 - 200 - 210 - 220 - 230 - 240 - 250 - 260
270 - 280 - 290 - 300 - 310 - 320 - 330 - 340 - 350 - 360 - 370 - 380 - 390
400 - 410 - 420 - 430 - 440 - 450 - 460 - 470 - 480 - 490 - 500 - 510 - 520
530 - 540 - 550 - 560 - 570 - 580 - 590 - 600 - 610 - 620 - 630 - 640 - 650
660 - 670 - 680 - 690 - 700 - 710 - 720 - 730 - 740 - 750 - 760 - 770 - 780
790 - 800 - 810 - 820 - 830 - 840 - 850 - 860 - 870 - 880 - 890 - 900 - 910
920 - 930 - 940 - 950 - 960 - 970 - 980 - 990 - 1000

Notes: ____

____

____

________________________________ **End Date:** ____________

Technique:______________________ **L - R - Alt. Start Date:** _______

Notes:____

____

____

____

Slow Reps: 10 - 20 - 30 - 40 - 50 - 60 - 70 - 80 - 90 - 100 - 110 - 120 - 130
140 - 150 - 160 - 170 - 180 - 190 - 200 - 210 - 220 - 230 - 240 - 250 - 260
270 - 280 - 290 - 300 - 310 - 320 - 330 - 340 - 350 - 360 - 370 - 380 - 390
400 - 410 - 420 - 430 - 440 - 450 - 460 - 470 - 480 - 490 - 500 - 510 - 520
530 - 540 - 550 - 560 - 570 - 580 - 590 - 600 - 610 - 620 - 630 - 640 - 650
660 - 670 - 680 - 690 - 700 - 710 - 720 - 730 - 740 - 750 - 760 - 770 - 780
790 - 800 - 810 - 820 - 830 - 840 - 850 - 860 - 870 - 880 - 890 - 900 - 910
920 - 930 - 940 - 950 - 960 - 970 - 980 - 990 - 1000

Reg Reps: 10 - 20 - 30 - 40 - 50 - 60 - 70 - 80 - 90 - 100 - 110 - 120 - 130
140 - 150 - 160 - 170 - 180 - 190 - 200 - 210 - 220 - 230 - 240 - 250 - 260
270 - 280 - 290 - 300 - 310 - 320 - 330 - 340 - 350 - 360 - 370 - 380 - 390
400 - 410 - 420 - 430 - 440 - 450 - 460 - 470 - 480 - 490 - 500 - 510 - 520
530 - 540 - 550 - 560 - 570 - 580 - 590 - 600 - 610 - 620 - 630 - 640 - 650
660 - 670 - 680 - 690 - 700 - 710 - 720 - 730 - 740 - 750 - 760 - 770 - 780
790 - 800 - 810 - 820 - 830 - 840 - 850 - 860 - 870 - 880 - 890 - 900 - 910
920 - 930 - 940 - 950 - 960 - 970 - 980 - 990 - 1000

Notes: ____

____

____

____

Reg Reps: 10 - 20 - 30 - 40 - 50 - 60 - 70 - 80 - 90 - 100 - 110 - 120 - 130
140 - 150 - 160 - 170 - 180 - 190 - 200 - 210 - 220 - 230 - 240 - 250 - 260
270 - 280 - 290 - 300 - 310 - 320 - 330 - 340 - 350 - 360 - 370 - 380 - 390
400 - 410 - 420 - 430 - 440 - 450 - 460 - 470 - 480 - 490 - 500 - 510 - 520
530 - 540 - 550 - 560 - 570 - 580 - 590 - 600 - 610 - 620 - 630 - 640 - 650
660 - 670 - 680 - 690 - 700 - 710 - 720 - 730 - 740 - 750 - 760 - 770 - 780
790 - 800 - 810 - 820 - 830 - 840 - 850 - 860 - 870 - 880 - 890 - 900 - 910
920 - 930 - 940 - 950 - 960 - 970 - 980 - 990 - 1000

Notes: ____

____

____

____

Reg Reps: 10 - 20 - 30 - 40 - 50 - 60 - 70 - 80 - 90 - 100 - 110 - 120 - 130
140 - 150 - 160 - 170 - 180 - 190 - 200 - 210 - 220 - 230 - 240 - 250 - 260
270 - 280 - 290 - 300 - 310 - 320 - 330 - 340 - 350 - 360 - 370 - 380 - 390
400 - 410 - 420 - 430 - 440 - 450 - 460 - 470 - 480 - 490 - 500 - 510 - 520
530 - 540 - 550 - 560 - 570 - 580 - 590 - 600 - 610 - 620 - 630 - 640 - 650
660 - 670 - 680 - 690 - 700 - 710 - 720 - 730 - 740 - 750 - 760 - 770 - 780
790 - 800 - 810 - 820 - 830 - 840 - 850 - 860 - 870 - 880 - 890 - 900 - 910
920 - 930 - 940 - 950 - 960 - 970 - 980 - 990 - 1000

Notes: ______________________________

____

____

____

Reg Reps: 10 - 20 - 30 - 40 - 50 - 60 - 70 - 80 - 90 - 100 - 110 - 120 - 130
140 - 150 - 160 - 170 - 180 - 190 - 200 - 210 - 220 - 230 - 240 - 250 - 260
270 - 280 - 290 - 300 - 310 - 320 - 330 - 340 - 350 - 360 - 370 - 380 - 390
400 - 410 - 420 - 430 - 440 - 450 - 460 - 470 - 480 - 490 - 500 - 510 - 520
530 - 540 - 550 - 560 - 570 - 580 - 590 - 600 - 610 - 620 - 630 - 640 - 650
660 - 670 - 680 - 690 - 700 - 710 - 720 - 730 - 740 - 750 - 760 - 770 - 780
790 - 800 - 810 - 820 - 830 - 840 - 850 - 860 - 870 - 880 - 890 - 900 - 910
920 - 930 - 940 - 950 - 960 - 970 - 980 - 990 - 1000

Notes: ______________________________

____

____

____

Reg Reps: 10 - 20 - 30 - 40 - 50 - 60 - 70 - 80 - 90 - 100 - 110 - 120 - 130
140 - 150 - 160 - 170 - 180 - 190 - 200 - 210 - 220 - 230 - 240 - 250 - 260
270 - 280 - 290 - 300 - 310 - 320 - 330 - 340 - 350 - 360 - 370 - 380 - 390
400 - 410 - 420 - 430 - 440 - 450 - 460 - 470 - 480 - 490 - 500 - 510 - 520
530 - 540 - 550 - 560 - 570 - 580 - 590 - 600 - 610 - 620 - 630 - 640 - 650
660 - 670 - 680 - 690 - 700 - 710 - 720 - 730 - 740 - 750 - 760 - 770 - 780
790 - 800 - 810 - 820 - 830 - 840 - 850 - 860 - 870 - 880 - 890 - 900 - 910
920 - 930 - 940 - 950 - 960 - 970 - 980 - 990 - 1000

Notes: ______________________________

____

____

______________________________ **End Date:** __________

Technique:__________________________ **L - R - Alt. Start Date:** ________
Notes:____

Slow Reps: 10 - 20 - 30 - 40 - 50 - 60 - 70 - 80 - 90 - 100 - 110 - 120 - 130
140 - 150 - 160 - 170 - 180 - 190 - 200 - 210 - 220 - 230 - 240 - 250 - 260
270 - 280 - 290 - 300 - 310 - 320 - 330 - 340 - 350 - 360 - 370 - 380 - 390
400 - 410 - 420 - 430 - 440 - 450 - 460 - 470 - 480 - 490 - 500 - 510 - 520
530 - 540 - 550 - 560 - 570 - 580 - 590 - 600 - 610 - 620 - 630 - 640 - 650
660 - 670 - 680 - 690 - 700 - 710 - 720 - 730 - 740 - 750 - 760 - 770 - 780
790 - 800 - 810 - 820 - 830 - 840 - 850 - 860 - 870 - 880 - 890 - 900 - 910
920 - 930 - 940 - 950 - 960 - 970 - 980 - 990 - 1000

Reg Reps: 10 - 20 - 30 - 40 - 50 - 60 - 70 - 80 - 90 - 100 - 110 - 120 - 130
140 - 150 - 160 - 170 - 180 - 190 - 200 - 210 - 220 - 230 - 240 - 250 - 260
270 - 280 - 290 - 300 - 310 - 320 - 330 - 340 - 350 - 360 - 370 - 380 - 390
400 - 410 - 420 - 430 - 440 - 450 - 460 - 470 - 480 - 490 - 500 - 510 - 520
530 - 540 - 550 - 560 - 570 - 580 - 590 - 600 - 610 - 620 - 630 - 640 - 650
660 - 670 - 680 - 690 - 700 - 710 - 720 - 730 - 740 - 750 - 760 - 770 - 780
790 - 800 - 810 - 820 - 830 - 840 - 850 - 860 - 870 - 880 - 890 - 900 - 910
920 - 930 - 940 - 950 - 960 - 970 - 980 - 990 - 1000
Notes: ____

Reg Reps: 10 - 20 - 30 - 40 - 50 - 60 - 70 - 80 - 90 - 100 - 110 - 120 - 130
140 - 150 - 160 - 170 - 180 - 190 - 200 - 210 - 220 - 230 - 240 - 250 - 260
270 - 280 - 290 - 300 - 310 - 320 - 330 - 340 - 350 - 360 - 370 - 380 - 390
400 - 410 - 420 - 430 - 440 - 450 - 460 - 470 - 480 - 490 - 500 - 510 - 520
530 - 540 - 550 - 560 - 570 - 580 - 590 - 600 - 610 - 620 - 630 - 640 - 650
660 - 670 - 680 - 690 - 700 - 710 - 720 - 730 - 740 - 750 - 760 - 770 - 780
790 - 800 - 810 - 820 - 830 - 840 - 850 - 860 - 870 - 880 - 890 - 900 - 910
920 - 930 - 940 - 950 - 960 - 970 - 980 - 990 - 1000
Notes: ____

Reg Reps: 10 - 20 - 30 - 40 - 50 - 60 - 70 - 80 - 90 - 100 - 110 - 120 - 130
140 - 150 - 160 - 170 - 180 - 190 - 200 - 210 - 220 - 230 - 240 - 250 - 260
270 - 280 - 290 - 300 - 310 - 320 - 330 - 340 - 350 - 360 - 370 - 380 - 390
400 - 410 - 420 - 430 - 440 - 450 - 460 - 470 - 480 - 490 - 500 - 510 - 520
530 - 540 - 550 - 560 - 570 - 580 - 590 - 600 - 610 - 620 - 630 - 640 - 650
660 - 670 - 680 - 690 - 700 - 710 - 720 - 730 - 740 - 750 - 760 - 770 - 780
790 - 800 - 810 - 820 - 830 - 840 - 850 - 860 - 870 - 880 - 890 - 900 - 910
920 - 930 - 940 - 950 - 960 - 970 - 980 - 990 - 1000

Notes: ____

____

____

____

Reg Reps: 10 - 20 - 30 - 40 - 50 - 60 - 70 - 80 - 90 - 100 - 110 - 120 - 130
140 - 150 - 160 - 170 - 180 - 190 - 200 - 210 - 220 - 230 - 240 - 250 - 260
270 - 280 - 290 - 300 - 310 - 320 - 330 - 340 - 350 - 360 - 370 - 380 - 390
400 - 410 - 420 - 430 - 440 - 450 - 460 - 470 - 480 - 490 - 500 - 510 - 520
530 - 540 - 550 - 560 - 570 - 580 - 590 - 600 - 610 - 620 - 630 - 640 - 650
660 - 670 - 680 - 690 - 700 - 710 - 720 - 730 - 740 - 750 - 760 - 770 - 780
790 - 800 - 810 - 820 - 830 - 840 - 850 - 860 - 870 - 880 - 890 - 900 - 910
920 - 930 - 940 - 950 - 960 - 970 - 980 - 990 - 1000

Notes: ____

____

____

____

Reg Reps: 10 - 20 - 30 - 40 - 50 - 60 - 70 - 80 - 90 - 100 - 110 - 120 - 130
140 - 150 - 160 - 170 - 180 - 190 - 200 - 210 - 220 - 230 - 240 - 250 - 260
270 - 280 - 290 - 300 - 310 - 320 - 330 - 340 - 350 - 360 - 370 - 380 - 390
400 - 410 - 420 - 430 - 440 - 450 - 460 - 470 - 480 - 490 - 500 - 510 - 520
530 - 540 - 550 - 560 - 570 - 580 - 590 - 600 - 610 - 620 - 630 - 640 - 650
660 - 670 - 680 - 690 - 700 - 710 - 720 - 730 - 740 - 750 - 760 - 770 - 780
790 - 800 - 810 - 820 - 830 - 840 - 850 - 860 - 870 - 880 - 890 - 900 - 910
920 - 930 - 940 - 950 - 960 - 970 - 980 - 990 - 1000

Notes: ____

____

____

______________________________ **End Date:** ____________

Technique: ______________________ **L - R - Alt. Start Date:** _______

Notes: ____

____

____

____

Slow Reps: 10 - 20 - 30 - 40 - 50 - 60 - 70 - 80 - 90 - 100 - 110 - 120 - 130
140 - 150 - 160 - 170 - 180 - 190 - 200 - 210 - 220 - 230 - 240 - 250 - 260
270 - 280 - 290 - 300 - 310 - 320 - 330 - 340 - 350 - 360 - 370 - 380 - 390
400 - 410 - 420 - 430 - 440 - 450 - 460 - 470 - 480 - 490 - 500 - 510 - 520
530 - 540 - 550 - 560 - 570 - 580 - 590 - 600 - 610 - 620 - 630 - 640 - 650
660 - 670 - 680 - 690 - 700 - 710 - 720 - 730 - 740 - 750 - 760 - 770 - 780
790 - 800 - 810 - 820 - 830 - 840 - 850 - 860 - 870 - 880 - 890 - 900 - 910
920 - 930 - 940 - 950 - 960 - 970 - 980 - 990 - 1000

Reg Reps: 10 - 20 - 30 - 40 - 50 - 60 - 70 - 80 - 90 - 100 - 110 - 120 - 130
140 - 150 - 160 - 170 - 180 - 190 - 200 - 210 - 220 - 230 - 240 - 250 - 260
270 - 280 - 290 - 300 - 310 - 320 - 330 - 340 - 350 - 360 - 370 - 380 - 390
400 - 410 - 420 - 430 - 440 - 450 - 460 - 470 - 480 - 490 - 500 - 510 - 520
530 - 540 - 550 - 560 - 570 - 580 - 590 - 600 - 610 - 620 - 630 - 640 - 650
660 - 670 - 680 - 690 - 700 - 710 - 720 - 730 - 740 - 750 - 760 - 770 - 780
790 - 800 - 810 - 820 - 830 - 840 - 850 - 860 - 870 - 880 - 890 - 900 - 910
920 - 930 - 940 - 950 - 960 - 970 - 980 - 990 - 1000

Notes: ____

____

____

____

Reg Reps: 10 - 20 - 30 - 40 - 50 - 60 - 70 - 80 - 90 - 100 - 110 - 120 - 130
140 - 150 - 160 - 170 - 180 - 190 - 200 - 210 - 220 - 230 - 240 - 250 - 260
270 - 280 - 290 - 300 - 310 - 320 - 330 - 340 - 350 - 360 - 370 - 380 - 390
400 - 410 - 420 - 430 - 440 - 450 - 460 - 470 - 480 - 490 - 500 - 510 - 520
530 - 540 - 550 - 560 - 570 - 580 - 590 - 600 - 610 - 620 - 630 - 640 - 650
660 - 670 - 680 - 690 - 700 - 710 - 720 - 730 - 740 - 750 - 760 - 770 - 780
790 - 800 - 810 - 820 - 830 - 840 - 850 - 860 - 870 - 880 - 890 - 900 - 910
920 - 930 - 940 - 950 - 960 - 970 - 980 - 990 - 1000

Notes: ____

____

____

____

Reg Reps: 10 - 20 - 30 - 40 - 50 - 60 - 70 - 80 - 90 - 100 - 110 - 120 - 130
140 - 150 - 160 - 170 - 180 - 190 - 200 - 210 - 220 - 230 - 240 - 250 - 260
270 - 280 - 290 - 300 - 310 - 320 - 330 - 340 - 350 - 360 - 370 - 380 - 390
400 - 410 - 420 - 430 - 440 - 450 - 460 - 470 - 480 - 490 - 500 - 510 - 520
530 - 540 - 550 - 560 - 570 - 580 - 590 - 600 - 610 - 620 - 630 - 640 - 650
660 - 670 - 680 - 690 - 700 - 710 - 720 - 730 - 740 - 750 - 760 - 770 - 780
790 - 800 - 810 - 820 - 830 - 840 - 850 - 860 - 870 - 880 - 890 - 900 - 910
920 - 930 - 940 - 950 - 960 - 970 - 980 - 990 - 1000

Notes: ____

____

____

____

Reg Reps: 10 - 20 - 30 - 40 - 50 - 60 - 70 - 80 - 90 - 100 - 110 - 120 - 130
140 - 150 - 160 - 170 - 180 - 190 - 200 - 210 - 220 - 230 - 240 - 250 - 260
270 - 280 - 290 - 300 - 310 - 320 - 330 - 340 - 350 - 360 - 370 - 380 - 390
400 - 410 - 420 - 430 - 440 - 450 - 460 - 470 - 480 - 490 - 500 - 510 - 520
530 - 540 - 550 - 560 - 570 - 580 - 590 - 600 - 610 - 620 - 630 - 640 - 650
660 - 670 - 680 - 690 - 700 - 710 - 720 - 730 - 740 - 750 - 760 - 770 - 780
790 - 800 - 810 - 820 - 830 - 840 - 850 - 860 - 870 - 880 - 890 - 900 - 910
920 - 930 - 940 - 950 - 960 - 970 - 980 - 990 - 1000

Notes: ____

____

____

____

Reg Reps: 10 - 20 - 30 - 40 - 50 - 60 - 70 - 80 - 90 - 100 - 110 - 120 - 130
140 - 150 - 160 - 170 - 180 - 190 - 200 - 210 - 220 - 230 - 240 - 250 - 260
270 - 280 - 290 - 300 - 310 - 320 - 330 - 340 - 350 - 360 - 370 - 380 - 390
400 - 410 - 420 - 430 - 440 - 450 - 460 - 470 - 480 - 490 - 500 - 510 - 520
530 - 540 - 550 - 560 - 570 - 580 - 590 - 600 - 610 - 620 - 630 - 640 - 650
660 - 670 - 680 - 690 - 700 - 710 - 720 - 730 - 740 - 750 - 760 - 770 - 780
790 - 800 - 810 - 820 - 830 - 840 - 850 - 860 - 870 - 880 - 890 - 900 - 910
920 - 930 - 940 - 950 - 960 - 970 - 980 - 990 - 1000

Notes: ____

____

____

________________________________ **End Date:** ____________

Technique:________________________ **L - R - Alt. Start Date:** _______

Notes:____

____

____

____

Slow Reps: 10 - 20 - 30 - 40 - 50 - 60 - 70 - 80 - 90 - 100 - 110 - 120 - 130
140 - 150 - 160 - 170 - 180 - 190 - 200 - 210 - 220 - 230 - 240 - 250 - 260
270 - 280 - 290 - 300 - 310 - 320 - 330 - 340 - 350 - 360 - 370 - 380 - 390
400 - 410 - 420 - 430 - 440 - 450 - 460 - 470 - 480 - 490 - 500 - 510 - 520
530 - 540 - 550 - 560 - 570 - 580 - 590 - 600 - 610 - 620 - 630 - 640 - 650
660 - 670 - 680 - 690 - 700 - 710 - 720 - 730 - 740 - 750 - 760 - 770 - 780
790 - 800 - 810 - 820 - 830 - 840 - 850 - 860 - 870 - 880 - 890 - 900 - 910
920 - 930 - 940 - 950 - 960 - 970 - 980 - 990 - 1000

Reg Reps: 10 - 20 - 30 - 40 - 50 - 60 - 70 - 80 - 90 - 100 - 110 - 120 - 130
140 - 150 - 160 - 170 - 180 - 190 - 200 - 210 - 220 - 230 - 240 - 250 - 260
270 - 280 - 290 - 300 - 310 - 320 - 330 - 340 - 350 - 360 - 370 - 380 - 390
400 - 410 - 420 - 430 - 440 - 450 - 460 - 470 - 480 - 490 - 500 - 510 - 520
530 - 540 - 550 - 560 - 570 - 580 - 590 - 600 - 610 - 620 - 630 - 640 - 650
660 - 670 - 680 - 690 - 700 - 710 - 720 - 730 - 740 - 750 - 760 - 770 - 780
790 - 800 - 810 - 820 - 830 - 840 - 850 - 860 - 870 - 880 - 890 - 900 - 910
920 - 930 - 940 - 950 - 960 - 970 - 980 - 990 - 1000

Notes: ____

____

____

____

Reg Reps: 10 - 20 - 30 - 40 - 50 - 60 - 70 - 80 - 90 - 100 - 110 - 120 - 130
140 - 150 - 160 - 170 - 180 - 190 - 200 - 210 - 220 - 230 - 240 - 250 - 260
270 - 280 - 290 - 300 - 310 - 320 - 330 - 340 - 350 - 360 - 370 - 380 - 390
400 - 410 - 420 - 430 - 440 - 450 - 460 - 470 - 480 - 490 - 500 - 510 - 520
530 - 540 - 550 - 560 - 570 - 580 - 590 - 600 - 610 - 620 - 630 - 640 - 650
660 - 670 - 680 - 690 - 700 - 710 - 720 - 730 - 740 - 750 - 760 - 770 - 780
790 - 800 - 810 - 820 - 830 - 840 - 850 - 860 - 870 - 880 - 890 - 900 - 910
920 - 930 - 940 - 950 - 960 - 970 - 980 - 990 - 1000

Notes: ____

____

____

____

Reg Reps: 10 - 20 - 30 - 40 - 50 - 60 - 70 - 80 - 90 - 100 - 110 - 120 - 130
140 - 150 - 160 - 170 - 180 - 190 - 200 - 210 - 220 - 230 - 240 - 250 - 260
270 - 280 - 290 - 300 - 310 - 320 - 330 - 340 - 350 - 360 - 370 - 380 - 390
400 - 410 - 420 - 430 - 440 - 450 - 460 - 470 - 480 - 490 - 500 - 510 - 520
530 - 540 - 550 - 560 - 570 - 580 - 590 - 600 - 610 - 620 - 630 - 640 - 650
660 - 670 - 680 - 690 - 700 - 710 - 720 - 730 - 740 - 750 - 760 - 770 - 780
790 - 800 - 810 - 820 - 830 - 840 - 850 - 860 - 870 - 880 - 890 - 900 - 910
920 - 930 - 940 - 950 - 960 - 970 - 980 - 990 - 1000

Notes: ____

____

____

____

Reg Reps: 10 - 20 - 30 - 40 - 50 - 60 - 70 - 80 - 90 - 100 - 110 - 120 - 130
140 - 150 - 160 - 170 - 180 - 190 - 200 - 210 - 220 - 230 - 240 - 250 - 260
270 - 280 - 290 - 300 - 310 - 320 - 330 - 340 - 350 - 360 - 370 - 380 - 390
400 - 410 - 420 - 430 - 440 - 450 - 460 - 470 - 480 - 490 - 500 - 510 - 520
530 - 540 - 550 - 560 - 570 - 580 - 590 - 600 - 610 - 620 - 630 - 640 - 650
660 - 670 - 680 - 690 - 700 - 710 - 720 - 730 - 740 - 750 - 760 - 770 - 780
790 - 800 - 810 - 820 - 830 - 840 - 850 - 860 - 870 - 880 - 890 - 900 - 910
920 - 930 - 940 - 950 - 960 - 970 - 980 - 990 - 1000

Notes: ____

____

____

____

Reg Reps: 10 - 20 - 30 - 40 - 50 - 60 - 70 - 80 - 90 - 100 - 110 - 120 - 130
140 - 150 - 160 - 170 - 180 - 190 - 200 - 210 - 220 - 230 - 240 - 250 - 260
270 - 280 - 290 - 300 - 310 - 320 - 330 - 340 - 350 - 360 - 370 - 380 - 390
400 - 410 - 420 - 430 - 440 - 450 - 460 - 470 - 480 - 490 - 500 - 510 - 520
530 - 540 - 550 - 560 - 570 - 580 - 590 - 600 - 610 - 620 - 630 - 640 - 650
660 - 670 - 680 - 690 - 700 - 710 - 720 - 730 - 740 - 750 - 760 - 770 - 780
790 - 800 - 810 - 820 - 830 - 840 - 850 - 860 - 870 - 880 - 890 - 900 - 910
920 - 930 - 940 - 950 - 960 - 970 - 980 - 990 - 1000

Notes: ____

____

____

__________________________________ **End Date:** __________

Technique:________________________ **L - R - Alt. Start Date:** _______

Notes:____

____

____

____

Slow Reps: 10 - 20 - 30 - 40 - 50 - 60 - 70 - 80 - 90 - 100 - 110 - 120 - 130
140 - 150 - 160 - 170 - 180 - 190 - 200 - 210 - 220 - 230 - 240 - 250 - 260
270 - 280 - 290 - 300 - 310 - 320 - 330 - 340 - 350 - 360 - 370 - 380 - 390
400 - 410 - 420 - 430 - 440 - 450 - 460 - 470 - 480 - 490 - 500 - 510 - 520
530 - 540 - 550 - 560 - 570 - 580 - 590 - 600 - 610 - 620 - 630 - 640 - 650
660 - 670 - 680 - 690 - 700 - 710 - 720 - 730 - 740 - 750 - 760 - 770 - 780
790 - 800 - 810 - 820 - 830 - 840 - 850 - 860 - 870 - 880 - 890 - 900 - 910
920 - 930 - 940 - 950 - 960 - 970 - 980 - 990 - 1000

Reg Reps: 10 - 20 - 30 - 40 - 50 - 60 - 70 - 80 - 90 - 100 - 110 - 120 - 130
140 - 150 - 160 - 170 - 180 - 190 - 200 - 210 - 220 - 230 - 240 - 250 - 260
270 - 280 - 290 - 300 - 310 - 320 - 330 - 340 - 350 - 360 - 370 - 380 - 390
400 - 410 - 420 - 430 - 440 - 450 - 460 - 470 - 480 - 490 - 500 - 510 - 520
530 - 540 - 550 - 560 - 570 - 580 - 590 - 600 - 610 - 620 - 630 - 640 - 650
660 - 670 - 680 - 690 - 700 - 710 - 720 - 730 - 740 - 750 - 760 - 770 - 780
790 - 800 - 810 - 820 - 830 - 840 - 850 - 860 - 870 - 880 - 890 - 900 - 910
920 - 930 - 940 - 950 - 960 - 970 - 980 - 990 - 1000

Notes: ____

____

____

____

Reg Reps: 10 - 20 - 30 - 40 - 50 - 60 - 70 - 80 - 90 - 100 - 110 - 120 - 130
140 - 150 - 160 - 170 - 180 - 190 - 200 - 210 - 220 - 230 - 240 - 250 - 260
270 - 280 - 290 - 300 - 310 - 320 - 330 - 340 - 350 - 360 - 370 - 380 - 390
400 - 410 - 420 - 430 - 440 - 450 - 460 - 470 - 480 - 490 - 500 - 510 - 520
530 - 540 - 550 - 560 - 570 - 580 - 590 - 600 - 610 - 620 - 630 - 640 - 650
660 - 670 - 680 - 690 - 700 - 710 - 720 - 730 - 740 - 750 - 760 - 770 - 780
790 - 800 - 810 - 820 - 830 - 840 - 850 - 860 - 870 - 880 - 890 - 900 - 910
920 - 930 - 940 - 950 - 960 - 970 - 980 - 990 - 1000

Notes: ____

____

____

____

Reg Reps: 10 - 20 - 30 - 40 - 50 - 60 - 70 - 80 - 90 - 100 - 110 - 120 - 130
140 - 150 - 160 - 170 - 180 - 190 - 200 - 210 - 220 - 230 - 240 - 250 - 260
270 - 280 - 290 - 300 - 310 - 320 - 330 - 340 - 350 - 360 - 370 - 380 - 390
400 - 410 - 420 - 430 - 440 - 450 - 460 - 470 - 480 - 490 - 500 - 510 - 520
530 - 540 - 550 - 560 - 570 - 580 - 590 - 600 - 610 - 620 - 630 - 640 - 650
660 - 670 - 680 - 690 - 700 - 710 - 720 - 730 - 740 - 750 - 760 - 770 - 780
790 - 800 - 810 - 820 - 830 - 840 - 850 - 860 - 870 - 880 - 890 - 900 - 910
920 - 930 - 940 - 950 - 960 - 970 - 980 - 990 - 1000

Notes: ____

____

____

____

Reg Reps: 10 - 20 - 30 - 40 - 50 - 60 - 70 - 80 - 90 - 100 - 110 - 120 - 130
140 - 150 - 160 - 170 - 180 - 190 - 200 - 210 - 220 - 230 - 240 - 250 - 260
270 - 280 - 290 - 300 - 310 - 320 - 330 - 340 - 350 - 360 - 370 - 380 - 390
400 - 410 - 420 - 430 - 440 - 450 - 460 - 470 - 480 - 490 - 500 - 510 - 520
530 - 540 - 550 - 560 - 570 - 580 - 590 - 600 - 610 - 620 - 630 - 640 - 650
660 - 670 - 680 - 690 - 700 - 710 - 720 - 730 - 740 - 750 - 760 - 770 - 780
790 - 800 - 810 - 820 - 830 - 840 - 850 - 860 - 870 - 880 - 890 - 900 - 910
920 - 930 - 940 - 950 - 960 - 970 - 980 - 990 - 1000

Notes: ____

____

____

____

Reg Reps: 10 - 20 - 30 - 40 - 50 - 60 - 70 - 80 - 90 - 100 - 110 - 120 - 130
140 - 150 - 160 - 170 - 180 - 190 - 200 - 210 - 220 - 230 - 240 - 250 - 260
270 - 280 - 290 - 300 - 310 - 320 - 330 - 340 - 350 - 360 - 370 - 380 - 390
400 - 410 - 420 - 430 - 440 - 450 - 460 - 470 - 480 - 490 - 500 - 510 - 520
530 - 540 - 550 - 560 - 570 - 580 - 590 - 600 - 610 - 620 - 630 - 640 - 650
660 - 670 - 680 - 690 - 700 - 710 - 720 - 730 - 740 - 750 - 760 - 770 - 780
790 - 800 - 810 - 820 - 830 - 840 - 850 - 860 - 870 - 880 - 890 - 900 - 910
920 - 930 - 940 - 950 - 960 - 970 - 980 - 990 - 1000

Notes: ____

____

____

______________________________ **End Date:** ____________

Technique:_________________________ **L - R - Alt. Start Date:** _______

Notes:____

____

____

____

Slow Reps: 10 - 20 - 30 - 40 - 50 - 60 - 70 - 80 - 90 - 100 - 110 - 120 - 130
140 - 150 - 160 - 170 - 180 - 190 - 200 - 210 - 220 - 230 - 240 - 250 - 260
270 - 280 - 290 - 300 - 310 - 320 - 330 - 340 - 350 - 360 - 370 - 380 - 390
400 - 410 - 420 - 430 - 440 - 450 - 460 - 470 - 480 - 490 - 500 - 510 - 520
530 - 540 - 550 - 560 - 570 - 580 - 590 - 600 - 610 - 620 - 630 - 640 - 650
660 - 670 - 680 - 690 - 700 - 710 - 720 - 730 - 740 - 750 - 760 - 770 - 780
790 - 800 - 810 - 820 - 830 - 840 - 850 - 860 - 870 - 880 - 890 - 900 - 910
920 - 930 - 940 - 950 - 960 - 970 - 980 - 990 - 1000

Reg Reps: 10 - 20 - 30 - 40 - 50 - 60 - 70 - 80 - 90 - 100 - 110 - 120 - 130
140 - 150 - 160 - 170 - 180 - 190 - 200 - 210 - 220 - 230 - 240 - 250 - 260
270 - 280 - 290 - 300 - 310 - 320 - 330 - 340 - 350 - 360 - 370 - 380 - 390
400 - 410 - 420 - 430 - 440 - 450 - 460 - 470 - 480 - 490 - 500 - 510 - 520
530 - 540 - 550 - 560 - 570 - 580 - 590 - 600 - 610 - 620 - 630 - 640 - 650
660 - 670 - 680 - 690 - 700 - 710 - 720 - 730 - 740 - 750 - 760 - 770 - 780
790 - 800 - 810 - 820 - 830 - 840 - 850 - 860 - 870 - 880 - 890 - 900 - 910
920 - 930 - 940 - 950 - 960 - 970 - 980 - 990 - 1000

Notes: ____

____

____

____

Reg Reps: 10 - 20 - 30 - 40 - 50 - 60 - 70 - 80 - 90 - 100 - 110 - 120 - 130
140 - 150 - 160 - 170 - 180 - 190 - 200 - 210 - 220 - 230 - 240 - 250 - 260
270 - 280 - 290 - 300 - 310 - 320 - 330 - 340 - 350 - 360 - 370 - 380 - 390
400 - 410 - 420 - 430 - 440 - 450 - 460 - 470 - 480 - 490 - 500 - 510 - 520
530 - 540 - 550 - 560 - 570 - 580 - 590 - 600 - 610 - 620 - 630 - 640 - 650
660 - 670 - 680 - 690 - 700 - 710 - 720 - 730 - 740 - 750 - 760 - 770 - 780
790 - 800 - 810 - 820 - 830 - 840 - 850 - 860 - 870 - 880 - 890 - 900 - 910
920 - 930 - 940 - 950 - 960 - 970 - 980 - 990 - 1000

Notes: ____

____

____

____

Reg Reps: 10 - 20 - 30 - 40 - 50 - 60 - 70 - 80 - 90 - 100 - 110 - 120 - 130
140 - 150 - 160 - 170 - 180 - 190 - 200 - 210 - 220 - 230 - 240 - 250 - 260
270 - 280 - 290 - 300 - 310 - 320 - 330 - 340 - 350 - 360 - 370 - 380 - 390
400 - 410 - 420 - 430 - 440 - 450 - 460 - 470 - 480 - 490 - 500 - 510 - 520
530 - 540 - 550 - 560 - 570 - 580 - 590 - 600 - 610 - 620 - 630 - 640 - 650
660 - 670 - 680 - 690 - 700 - 710 - 720 - 730 - 740 - 750 - 760 - 770 - 780
790 - 800 - 810 - 820 - 830 - 840 - 850 - 860 - 870 - 880 - 890 - 900 - 910
920 - 930 - 940 - 950 - 960 - 970 - 980 - 990 - 1000

Notes: ____

____

____

____

Reg Reps: 10 - 20 - 30 - 40 - 50 - 60 - 70 - 80 - 90 - 100 - 110 - 120 - 130
140 - 150 - 160 - 170 - 180 - 190 - 200 - 210 - 220 - 230 - 240 - 250 - 260
270 - 280 - 290 - 300 - 310 - 320 - 330 - 340 - 350 - 360 - 370 - 380 - 390
400 - 410 - 420 - 430 - 440 - 450 - 460 - 470 - 480 - 490 - 500 - 510 - 520
530 - 540 - 550 - 560 - 570 - 580 - 590 - 600 - 610 - 620 - 630 - 640 - 650
660 - 670 - 680 - 690 - 700 - 710 - 720 - 730 - 740 - 750 - 760 - 770 - 780
790 - 800 - 810 - 820 - 830 - 840 - 850 - 860 - 870 - 880 - 890 - 900 - 910
920 - 930 - 940 - 950 - 960 - 970 - 980 - 990 - 1000

Notes: ____

____

____

____

Reg Reps: 10 - 20 - 30 - 40 - 50 - 60 - 70 - 80 - 90 - 100 - 110 - 120 - 130
140 - 150 - 160 - 170 - 180 - 190 - 200 - 210 - 220 - 230 - 240 - 250 - 260
270 - 280 - 290 - 300 - 310 - 320 - 330 - 340 - 350 - 360 - 370 - 380 - 390
400 - 410 - 420 - 430 - 440 - 450 - 460 - 470 - 480 - 490 - 500 - 510 - 520
530 - 540 - 550 - 560 - 570 - 580 - 590 - 600 - 610 - 620 - 630 - 640 - 650
660 - 670 - 680 - 690 - 700 - 710 - 720 - 730 - 740 - 750 - 760 - 770 - 780
790 - 800 - 810 - 820 - 830 - 840 - 850 - 860 - 870 - 880 - 890 - 900 - 910
920 - 930 - 940 - 950 - 960 - 970 - 980 - 990 - 1000

Notes: ____

____

____

______________________________ **End Date:** ____________

Technique: ________________________ **L - R - Alt. Start Date:** _______

Notes: ____

____

____

____

Slow Reps: 10 - 20 - 30 - 40 - 50 - 60 - 70 - 80 - 90 - 100 - 110 - 120 - 130
140 - 150 - 160 - 170 - 180 - 190 - 200 - 210 - 220 - 230 - 240 - 250 - 260
270 - 280 - 290 - 300 - 310 - 320 - 330 - 340 - 350 - 360 - 370 - 380 - 390
400 - 410 - 420 - 430 - 440 - 450 - 460 - 470 - 480 - 490 - 500 - 510 - 520
530 - 540 - 550 - 560 - 570 - 580 - 590 - 600 - 610 - 620 - 630 - 640 - 650
660 - 670 - 680 - 690 - 700 - 710 - 720 - 730 - 740 - 750 - 760 - 770 - 780
790 - 800 - 810 - 820 - 830 - 840 - 850 - 860 - 870 - 880 - 890 - 900 - 910
920 - 930 - 940 - 950 - 960 - 970 - 980 - 990 - 1000

Reg Reps: 10 - 20 - 30 - 40 - 50 - 60 - 70 - 80 - 90 - 100 - 110 - 120 - 130
140 - 150 - 160 - 170 - 180 - 190 - 200 - 210 - 220 - 230 - 240 - 250 - 260
270 - 280 - 290 - 300 - 310 - 320 - 330 - 340 - 350 - 360 - 370 - 380 - 390
400 - 410 - 420 - 430 - 440 - 450 - 460 - 470 - 480 - 490 - 500 - 510 - 520
530 - 540 - 550 - 560 - 570 - 580 - 590 - 600 - 610 - 620 - 630 - 640 - 650
660 - 670 - 680 - 690 - 700 - 710 - 720 - 730 - 740 - 750 - 760 - 770 - 780
790 - 800 - 810 - 820 - 830 - 840 - 850 - 860 - 870 - 880 - 890 - 900 - 910
920 - 930 - 940 - 950 - 960 - 970 - 980 - 990 - 1000

Notes: ____

____

____

____

Reg Reps: 10 - 20 - 30 - 40 - 50 - 60 - 70 - 80 - 90 - 100 - 110 - 120 - 130
140 - 150 - 160 - 170 - 180 - 190 - 200 - 210 - 220 - 230 - 240 - 250 - 260
270 - 280 - 290 - 300 - 310 - 320 - 330 - 340 - 350 - 360 - 370 - 380 - 390
400 - 410 - 420 - 430 - 440 - 450 - 460 - 470 - 480 - 490 - 500 - 510 - 520
530 - 540 - 550 - 560 - 570 - 580 - 590 - 600 - 610 - 620 - 630 - 640 - 650
660 - 670 - 680 - 690 - 700 - 710 - 720 - 730 - 740 - 750 - 760 - 770 - 780
790 - 800 - 810 - 820 - 830 - 840 - 850 - 860 - 870 - 880 - 890 - 900 - 910
920 - 930 - 940 - 950 - 960 - 970 - 980 - 990 - 1000

Notes: ____

____

____

____

Reg Reps: 10 - 20 - 30 - 40 - 50 - 60 - 70 - 80 - 90 - 100 - 110 - 120 - 130
140 - 150 - 160 - 170 - 180 - 190 - 200 - 210 - 220 - 230 - 240 - 250 - 260
270 - 280 - 290 - 300 - 310 - 320 - 330 - 340 - 350 - 360 - 370 - 380 - 390
400 - 410 - 420 - 430 - 440 - 450 - 460 - 470 - 480 - 490 - 500 - 510 - 520
530 - 540 - 550 - 560 - 570 - 580 - 590 - 600 - 610 - 620 - 630 - 640 - 650
660 - 670 - 680 - 690 - 700 - 710 - 720 - 730 - 740 - 750 - 760 - 770 - 780
790 - 800 - 810 - 820 - 830 - 840 - 850 - 860 - 870 - 880 - 890 - 900 - 910
920 - 930 - 940 - 950 - 960 - 970 - 980 - 990 - 1000

Notes: ____

____

____

____

Reg Reps: 10 - 20 - 30 - 40 - 50 - 60 - 70 - 80 - 90 - 100 - 110 - 120 - 130
140 - 150 - 160 - 170 - 180 - 190 - 200 - 210 - 220 - 230 - 240 - 250 - 260
270 - 280 - 290 - 300 - 310 - 320 - 330 - 340 - 350 - 360 - 370 - 380 - 390
400 - 410 - 420 - 430 - 440 - 450 - 460 - 470 - 480 - 490 - 500 - 510 - 520
530 - 540 - 550 - 560 - 570 - 580 - 590 - 600 - 610 - 620 - 630 - 640 - 650
660 - 670 - 680 - 690 - 700 - 710 - 720 - 730 - 740 - 750 - 760 - 770 - 780
790 - 800 - 810 - 820 - 830 - 840 - 850 - 860 - 870 - 880 - 890 - 900 - 910
920 - 930 - 940 - 950 - 960 - 970 - 980 - 990 - 1000

Notes: ____

____

____

____

Reg Reps: 10 - 20 - 30 - 40 - 50 - 60 - 70 - 80 - 90 - 100 - 110 - 120 - 130
140 - 150 - 160 - 170 - 180 - 190 - 200 - 210 - 220 - 230 - 240 - 250 - 260
270 - 280 - 290 - 300 - 310 - 320 - 330 - 340 - 350 - 360 - 370 - 380 - 390
400 - 410 - 420 - 430 - 440 - 450 - 460 - 470 - 480 - 490 - 500 - 510 - 520
530 - 540 - 550 - 560 - 570 - 580 - 590 - 600 - 610 - 620 - 630 - 640 - 650
660 - 670 - 680 - 690 - 700 - 710 - 720 - 730 - 740 - 750 - 760 - 770 - 780
790 - 800 - 810 - 820 - 830 - 840 - 850 - 860 - 870 - 880 - 890 - 900 - 910
920 - 930 - 940 - 950 - 960 - 970 - 980 - 990 - 1000

Notes: ____

____

____

____________________________________ **End Date:** __________

Technique:____________________ **L - R - Alt. Start Date:** _______

Notes:____

Slow Reps: 10 - 20 - 30 - 40 - 50 - 60 - 70 - 80 - 90 - 100 - 110 - 120 - 130
140 - 150 - 160 - 170 - 180 - 190 - 200 - 210 - 220 - 230 - 240 - 250 - 260
270 - 280 - 290 - 300 - 310 - 320 - 330 - 340 - 350 - 360 - 370 - 380 - 390
400 - 410 - 420 - 430 - 440 - 450 - 460 - 470 - 480 - 490 - 500 - 510 - 520
530 - 540 - 550 - 560 - 570 - 580 - 590 - 600 - 610 - 620 - 630 - 640 - 650
660 - 670 - 680 - 690 - 700 - 710 - 720 - 730 - 740 - 750 - 760 - 770 - 780
790 - 800 - 810 - 820 - 830 - 840 - 850 - 860 - 870 - 880 - 890 - 900 - 910
920 - 930 - 940 - 950 - 960 - 970 - 980 - 990 - 1000

Reg Reps: 10 - 20 - 30 - 40 - 50 - 60 - 70 - 80 - 90 - 100 - 110 - 120 - 130
140 - 150 - 160 - 170 - 180 - 190 - 200 - 210 - 220 - 230 - 240 - 250 - 260
270 - 280 - 290 - 300 - 310 - 320 - 330 - 340 - 350 - 360 - 370 - 380 - 390
400 - 410 - 420 - 430 - 440 - 450 - 460 - 470 - 480 - 490 - 500 - 510 - 520
530 - 540 - 550 - 560 - 570 - 580 - 590 - 600 - 610 - 620 - 630 - 640 - 650
660 - 670 - 680 - 690 - 700 - 710 - 720 - 730 - 740 - 750 - 760 - 770 - 780
790 - 800 - 810 - 820 - 830 - 840 - 850 - 860 - 870 - 880 - 890 - 900 - 910
920 - 930 - 940 - 950 - 960 - 970 - 980 - 990 - 1000

Notes: ____

Reg Reps: 10 - 20 - 30 - 40 - 50 - 60 - 70 - 80 - 90 - 100 - 110 - 120 - 130
140 - 150 - 160 - 170 - 180 - 190 - 200 - 210 - 220 - 230 - 240 - 250 - 260
270 - 280 - 290 - 300 - 310 - 320 - 330 - 340 - 350 - 360 - 370 - 380 - 390
400 - 410 - 420 - 430 - 440 - 450 - 460 - 470 - 480 - 490 - 500 - 510 - 520
530 - 540 - 550 - 560 - 570 - 580 - 590 - 600 - 610 - 620 - 630 - 640 - 650
660 - 670 - 680 - 690 - 700 - 710 - 720 - 730 - 740 - 750 - 760 - 770 - 780
790 - 800 - 810 - 820 - 830 - 840 - 850 - 860 - 870 - 880 - 890 - 900 - 910
920 - 930 - 940 - 950 - 960 - 970 - 980 - 990 - 1000

Notes: ____

Reg Reps: 10 - 20 - 30 - 40 - 50 - 60 - 70 - 80 - 90 - 100 - 110 - 120 - 130
140 - 150 - 160 - 170 - 180 - 190 - 200 - 210 - 220 - 230 - 240 - 250 - 260
270 - 280 - 290 - 300 - 310 - 320 - 330 - 340 - 350 - 360 - 370 - 380 - 390
400 - 410 - 420 - 430 - 440 - 450 - 460 - 470 - 480 - 490 - 500 - 510 - 520
530 - 540 - 550 - 560 - 570 - 580 - 590 - 600 - 610 - 620 - 630 - 640 - 650
660 - 670 - 680 - 690 - 700 - 710 - 720 - 730 - 740 - 750 - 760 - 770 - 780
790 - 800 - 810 - 820 - 830 - 840 - 850 - 860 - 870 - 880 - 890 - 900 - 910
920 - 930 - 940 - 950 - 960 - 970 - 980 - 990 - 1000

Notes: ______________________________

____

____

____

Reg Reps: 10 - 20 - 30 - 40 - 50 - 60 - 70 - 80 - 90 - 100 - 110 - 120 - 130
140 - 150 - 160 - 170 - 180 - 190 - 200 - 210 - 220 - 230 - 240 - 250 - 260
270 - 280 - 290 - 300 - 310 - 320 - 330 - 340 - 350 - 360 - 370 - 380 - 390
400 - 410 - 420 - 430 - 440 - 450 - 460 - 470 - 480 - 490 - 500 - 510 - 520
530 - 540 - 550 - 560 - 570 - 580 - 590 - 600 - 610 - 620 - 630 - 640 - 650
660 - 670 - 680 - 690 - 700 - 710 - 720 - 730 - 740 - 750 - 760 - 770 - 780
790 - 800 - 810 - 820 - 830 - 840 - 850 - 860 - 870 - 880 - 890 - 900 - 910
920 - 930 - 940 - 950 - 960 - 970 - 980 - 990 - 1000

Notes: ______________________________

____

____

____

Reg Reps: 10 - 20 - 30 - 40 - 50 - 60 - 70 - 80 - 90 - 100 - 110 - 120 - 130
140 - 150 - 160 - 170 - 180 - 190 - 200 - 210 - 220 - 230 - 240 - 250 - 260
270 - 280 - 290 - 300 - 310 - 320 - 330 - 340 - 350 - 360 - 370 - 380 - 390
400 - 410 - 420 - 430 - 440 - 450 - 460 - 470 - 480 - 490 - 500 - 510 - 520
530 - 540 - 550 - 560 - 570 - 580 - 590 - 600 - 610 - 620 - 630 - 640 - 650
660 - 670 - 680 - 690 - 700 - 710 - 720 - 730 - 740 - 750 - 760 - 770 - 780
790 - 800 - 810 - 820 - 830 - 840 - 850 - 860 - 870 - 880 - 890 - 900 - 910
920 - 930 - 940 - 950 - 960 - 970 - 980 - 990 - 1000

Notes: ______________________________

____

____

______________________________ **End Date:** ____________

Technique: ______________________ **L - R - Alt. Start Date:** ______

Notes: ____

____

____

____

Slow Reps: 10 - 20 - 30 - 40 - 50 - 60 - 70 - 80 - 90 - 100 - 110 - 120 - 130
140 - 150 - 160 - 170 - 180 - 190 - 200 - 210 - 220 - 230 - 240 - 250 - 260
270 - 280 - 290 - 300 - 310 - 320 - 330 - 340 - 350 - 360 - 370 - 380 - 390
400 - 410 - 420 - 430 - 440 - 450 - 460 - 470 - 480 - 490 - 500 - 510 - 520
530 - 540 - 550 - 560 - 570 - 580 - 590 - 600 - 610 - 620 - 630 - 640 - 650
660 - 670 - 680 - 690 - 700 - 710 - 720 - 730 - 740 - 750 - 760 - 770 - 780
790 - 800 - 810 - 820 - 830 - 840 - 850 - 860 - 870 - 880 - 890 - 900 - 910
920 - 930 - 940 - 950 - 960 - 970 - 980 - 990 - 1000

Reg Reps: 10 - 20 - 30 - 40 - 50 - 60 - 70 - 80 - 90 - 100 - 110 - 120 - 130
140 - 150 - 160 - 170 - 180 - 190 - 200 - 210 - 220 - 230 - 240 - 250 - 260
270 - 280 - 290 - 300 - 310 - 320 - 330 - 340 - 350 - 360 - 370 - 380 - 390
400 - 410 - 420 - 430 - 440 - 450 - 460 - 470 - 480 - 490 - 500 - 510 - 520
530 - 540 - 550 - 560 - 570 - 580 - 590 - 600 - 610 - 620 - 630 - 640 - 650
660 - 670 - 680 - 690 - 700 - 710 - 720 - 730 - 740 - 750 - 760 - 770 - 780
790 - 800 - 810 - 820 - 830 - 840 - 850 - 860 - 870 - 880 - 890 - 900 - 910
920 - 930 - 940 - 950 - 960 - 970 - 980 - 990 - 1000

Notes: ____

____

____

____

Reg Reps: 10 - 20 - 30 - 40 - 50 - 60 - 70 - 80 - 90 - 100 - 110 - 120 - 130
140 - 150 - 160 - 170 - 180 - 190 - 200 - 210 - 220 - 230 - 240 - 250 - 260
270 - 280 - 290 - 300 - 310 - 320 - 330 - 340 - 350 - 360 - 370 - 380 - 390
400 - 410 - 420 - 430 - 440 - 450 - 460 - 470 - 480 - 490 - 500 - 510 - 520
530 - 540 - 550 - 560 - 570 - 580 - 590 - 600 - 610 - 620 - 630 - 640 - 650
660 - 670 - 680 - 690 - 700 - 710 - 720 - 730 - 740 - 750 - 760 - 770 - 780
790 - 800 - 810 - 820 - 830 - 840 - 850 - 860 - 870 - 880 - 890 - 900 - 910
920 - 930 - 940 - 950 - 960 - 970 - 980 - 990 - 1000

Notes: ____

____

____

____

Reg Reps: 10 - 20 - 30 - 40 - 50 - 60 - 70 - 80 - 90 - 100 - 110 - 120 - 130
140 - 150 - 160 - 170 - 180 - 190 - 200 - 210 - 220 - 230 - 240 - 250 - 260
270 - 280 - 290 - 300 - 310 - 320 - 330 - 340 - 350 - 360 - 370 - 380 - 390
400 - 410 - 420 - 430 - 440 - 450 - 460 - 470 - 480 - 490 - 500 - 510 - 520
530 - 540 - 550 - 560 - 570 - 580 - 590 - 600 - 610 - 620 - 630 - 640 - 650
660 - 670 - 680 - 690 - 700 - 710 - 720 - 730 - 740 - 750 - 760 - 770 - 780
790 - 800 - 810 - 820 - 830 - 840 - 850 - 860 - 870 - 880 - 890 - 900 - 910
920 - 930 - 940 - 950 - 960 - 970 - 980 - 990 - 1000

Notes: ______________________________

____

____

____

Reg Reps: 10 - 20 - 30 - 40 - 50 - 60 - 70 - 80 - 90 - 100 - 110 - 120 - 130
140 - 150 - 160 - 170 - 180 - 190 - 200 - 210 - 220 - 230 - 240 - 250 - 260
270 - 280 - 290 - 300 - 310 - 320 - 330 - 340 - 350 - 360 - 370 - 380 - 390
400 - 410 - 420 - 430 - 440 - 450 - 460 - 470 - 480 - 490 - 500 - 510 - 520
530 - 540 - 550 - 560 - 570 - 580 - 590 - 600 - 610 - 620 - 630 - 640 - 650
660 - 670 - 680 - 690 - 700 - 710 - 720 - 730 - 740 - 750 - 760 - 770 - 780
790 - 800 - 810 - 820 - 830 - 840 - 850 - 860 - 870 - 880 - 890 - 900 - 910
920 - 930 - 940 - 950 - 960 - 970 - 980 - 990 - 1000

Notes: ______________________________

____

____

____

Reg Reps: 10 - 20 - 30 - 40 - 50 - 60 - 70 - 80 - 90 - 100 - 110 - 120 - 130
140 - 150 - 160 - 170 - 180 - 190 - 200 - 210 - 220 - 230 - 240 - 250 - 260
270 - 280 - 290 - 300 - 310 - 320 - 330 - 340 - 350 - 360 - 370 - 380 - 390
400 - 410 - 420 - 430 - 440 - 450 - 460 - 470 - 480 - 490 - 500 - 510 - 520
530 - 540 - 550 - 560 - 570 - 580 - 590 - 600 - 610 - 620 - 630 - 640 - 650
660 - 670 - 680 - 690 - 700 - 710 - 720 - 730 - 740 - 750 - 760 - 770 - 780
790 - 800 - 810 - 820 - 830 - 840 - 850 - 860 - 870 - 880 - 890 - 900 - 910
920 - 930 - 940 - 950 - 960 - 970 - 980 - 990 - 1000

Notes: ______________________________

____

____

______________________________ **End Date:** ____________

Technique:_______________________ **L - R - Alt. Start Date:** _______

Notes:____

____

____

____

Slow Reps: 10 - 20 - 30 - 40 - 50 - 60 - 70 - 80 - 90 - 100 - 110 - 120 - 130
140 - 150 - 160 - 170 - 180 - 190 - 200 - 210 - 220 - 230 - 240 - 250 - 260
270 - 280 - 290 - 300 - 310 - 320 - 330 - 340 - 350 - 360 - 370 - 380 - 390
400 - 410 - 420 - 430 - 440 - 450 - 460 - 470 - 480 - 490 - 500 - 510 - 520
530 - 540 - 550 - 560 - 570 - 580 - 590 - 600 - 610 - 620 - 630 - 640 - 650
660 - 670 - 680 - 690 - 700 - 710 - 720 - 730 - 740 - 750 - 760 - 770 - 780
790 - 800 - 810 - 820 - 830 - 840 - 850 - 860 - 870 - 880 - 890 - 900 - 910
920 - 930 - 940 - 950 - 960 - 970 - 980 - 990 - 1000

Reg Reps: 10 - 20 - 30 - 40 - 50 - 60 - 70 - 80 - 90 - 100 - 110 - 120 - 130
140 - 150 - 160 - 170 - 180 - 190 - 200 - 210 - 220 - 230 - 240 - 250 - 260
270 - 280 - 290 - 300 - 310 - 320 - 330 - 340 - 350 - 360 - 370 - 380 - 390
400 - 410 - 420 - 430 - 440 - 450 - 460 - 470 - 480 - 490 - 500 - 510 - 520
530 - 540 - 550 - 560 - 570 - 580 - 590 - 600 - 610 - 620 - 630 - 640 - 650
660 - 670 - 680 - 690 - 700 - 710 - 720 - 730 - 740 - 750 - 760 - 770 - 780
790 - 800 - 810 - 820 - 830 - 840 - 850 - 860 - 870 - 880 - 890 - 900 - 910
920 - 930 - 940 - 950 - 960 - 970 - 980 - 990 - 1000

Notes: ____

____

____

____

Reg Reps: 10 - 20 - 30 - 40 - 50 - 60 - 70 - 80 - 90 - 100 - 110 - 120 - 130
140 - 150 - 160 - 170 - 180 - 190 - 200 - 210 - 220 - 230 - 240 - 250 - 260
270 - 280 - 290 - 300 - 310 - 320 - 330 - 340 - 350 - 360 - 370 - 380 - 390
400 - 410 - 420 - 430 - 440 - 450 - 460 - 470 - 480 - 490 - 500 - 510 - 520
530 - 540 - 550 - 560 - 570 - 580 - 590 - 600 - 610 - 620 - 630 - 640 - 650
660 - 670 - 680 - 690 - 700 - 710 - 720 - 730 - 740 - 750 - 760 - 770 - 780
790 - 800 - 810 - 820 - 830 - 840 - 850 - 860 - 870 - 880 - 890 - 900 - 910
920 - 930 - 940 - 950 - 960 - 970 - 980 - 990 - 1000

Notes: ____

____

____

____

Reg Reps: 10 - 20 - 30 - 40 - 50 - 60 - 70 - 80 - 90 - 100 - 110 - 120 - 130
140 - 150 - 160 - 170 - 180 - 190 - 200 - 210 - 220 - 230 - 240 - 250 - 260
270 - 280 - 290 - 300 - 310 - 320 - 330 - 340 - 350 - 360 - 370 - 380 - 390
400 - 410 - 420 - 430 - 440 - 450 - 460 - 470 - 480 - 490 - 500 - 510 - 520
530 - 540 - 550 - 560 - 570 - 580 - 590 - 600 - 610 - 620 - 630 - 640 - 650
660 - 670 - 680 - 690 - 700 - 710 - 720 - 730 - 740 - 750 - 760 - 770 - 780
790 - 800 - 810 - 820 - 830 - 840 - 850 - 860 - 870 - 880 - 890 - 900 - 910
920 - 930 - 940 - 950 - 960 - 970 - 980 - 990 - 1000

Notes: ____

____

____

____

Reg Reps: 10 - 20 - 30 - 40 - 50 - 60 - 70 - 80 - 90 - 100 - 110 - 120 - 130
140 - 150 - 160 - 170 - 180 - 190 - 200 - 210 - 220 - 230 - 240 - 250 - 260
270 - 280 - 290 - 300 - 310 - 320 - 330 - 340 - 350 - 360 - 370 - 380 - 390
400 - 410 - 420 - 430 - 440 - 450 - 460 - 470 - 480 - 490 - 500 - 510 - 520
530 - 540 - 550 - 560 - 570 - 580 - 590 - 600 - 610 - 620 - 630 - 640 - 650
660 - 670 - 680 - 690 - 700 - 710 - 720 - 730 - 740 - 750 - 760 - 770 - 780
790 - 800 - 810 - 820 - 830 - 840 - 850 - 860 - 870 - 880 - 890 - 900 - 910
920 - 930 - 940 - 950 - 960 - 970 - 980 - 990 - 1000

Notes: ____

____

____

____

Reg Reps: 10 - 20 - 30 - 40 - 50 - 60 - 70 - 80 - 90 - 100 - 110 - 120 - 130
140 - 150 - 160 - 170 - 180 - 190 - 200 - 210 - 220 - 230 - 240 - 250 - 260
270 - 280 - 290 - 300 - 310 - 320 - 330 - 340 - 350 - 360 - 370 - 380 - 390
400 - 410 - 420 - 430 - 440 - 450 - 460 - 470 - 480 - 490 - 500 - 510 - 520
530 - 540 - 550 - 560 - 570 - 580 - 590 - 600 - 610 - 620 - 630 - 640 - 650
660 - 670 - 680 - 690 - 700 - 710 - 720 - 730 - 740 - 750 - 760 - 770 - 780
790 - 800 - 810 - 820 - 830 - 840 - 850 - 860 - 870 - 880 - 890 - 900 - 910
920 - 930 - 940 - 950 - 960 - 970 - 980 - 990 - 1000

Notes: ____

____

____

______________________________ **End Date:** ____________

Technique:________________________ **L - R - Alt. Start Date:** _______

Notes:____

Slow Reps: 10 - 20 - 30 - 40 - 50 - 60 - 70 - 80 - 90 - 100 - 110 - 120 - 130
140 - 150 - 160 - 170 - 180 - 190 - 200 - 210 - 220 - 230 - 240 - 250 - 260
270 - 280 - 290 - 300 - 310 - 320 - 330 - 340 - 350 - 360 - 370 - 380 - 390
400 - 410 - 420 - 430 - 440 - 450 - 460 - 470 - 480 - 490 - 500 - 510 - 520
530 - 540 - 550 - 560 - 570 - 580 - 590 - 600 - 610 - 620 - 630 - 640 - 650
660 - 670 - 680 - 690 - 700 - 710 - 720 - 730 - 740 - 750 - 760 - 770 - 780
790 - 800 - 810 - 820 - 830 - 840 - 850 - 860 - 870 - 880 - 890 - 900 - 910
920 - 930 - 940 - 950 - 960 - 970 - 980 - 990 - 1000

Reg Reps: 10 - 20 - 30 - 40 - 50 - 60 - 70 - 80 - 90 - 100 - 110 - 120 - 130
140 - 150 - 160 - 170 - 180 - 190 - 200 - 210 - 220 - 230 - 240 - 250 - 260
270 - 280 - 290 - 300 - 310 - 320 - 330 - 340 - 350 - 360 - 370 - 380 - 390
400 - 410 - 420 - 430 - 440 - 450 - 460 - 470 - 480 - 490 - 500 - 510 - 520
530 - 540 - 550 - 560 - 570 - 580 - 590 - 600 - 610 - 620 - 630 - 640 - 650
660 - 670 - 680 - 690 - 700 - 710 - 720 - 730 - 740 - 750 - 760 - 770 - 780
790 - 800 - 810 - 820 - 830 - 840 - 850 - 860 - 870 - 880 - 890 - 900 - 910
920 - 930 - 940 - 950 - 960 - 970 - 980 - 990 - 1000

Notes: ____

Reg Reps: 10 - 20 - 30 - 40 - 50 - 60 - 70 - 80 - 90 - 100 - 110 - 120 - 130
140 - 150 - 160 - 170 - 180 - 190 - 200 - 210 - 220 - 230 - 240 - 250 - 260
270 - 280 - 290 - 300 - 310 - 320 - 330 - 340 - 350 - 360 - 370 - 380 - 390
400 - 410 - 420 - 430 - 440 - 450 - 460 - 470 - 480 - 490 - 500 - 510 - 520
530 - 540 - 550 - 560 - 570 - 580 - 590 - 600 - 610 - 620 - 630 - 640 - 650
660 - 670 - 680 - 690 - 700 - 710 - 720 - 730 - 740 - 750 - 760 - 770 - 780
790 - 800 - 810 - 820 - 830 - 840 - 850 - 860 - 870 - 880 - 890 - 900 - 910
920 - 930 - 940 - 950 - 960 - 970 - 980 - 990 - 1000

Notes: ____

Reg Reps: 10 - 20 - 30 - 40 - 50 - 60 - 70 - 80 - 90 - 100 - 110 - 120 - 130
140 - 150 - 160 - 170 - 180 - 190 - 200 - 210 - 220 - 230 - 240 - 250 - 260
270 - 280 - 290 - 300 - 310 - 320 - 330 - 340 - 350 - 360 - 370 - 380 - 390
400 - 410 - 420 - 430 - 440 - 450 - 460 - 470 - 480 - 490 - 500 - 510 - 520
530 - 540 - 550 - 560 - 570 - 580 - 590 - 600 - 610 - 620 - 630 - 640 - 650
660 - 670 - 680 - 690 - 700 - 710 - 720 - 730 - 740 - 750 - 760 - 770 - 780
790 - 800 - 810 - 820 - 830 - 840 - 850 - 860 - 870 - 880 - 890 - 900 - 910
920 - 930 - 940 - 950 - 960 - 970 - 980 - 990 - 1000

Notes: ____

____

____

____

Reg Reps: 10 - 20 - 30 - 40 - 50 - 60 - 70 - 80 - 90 - 100 - 110 - 120 - 130
140 - 150 - 160 - 170 - 180 - 190 - 200 - 210 - 220 - 230 - 240 - 250 - 260
270 - 280 - 290 - 300 - 310 - 320 - 330 - 340 - 350 - 360 - 370 - 380 - 390
400 - 410 - 420 - 430 - 440 - 450 - 460 - 470 - 480 - 490 - 500 - 510 - 520
530 - 540 - 550 - 560 - 570 - 580 - 590 - 600 - 610 - 620 - 630 - 640 - 650
660 - 670 - 680 - 690 - 700 - 710 - 720 - 730 - 740 - 750 - 760 - 770 - 780
790 - 800 - 810 - 820 - 830 - 840 - 850 - 860 - 870 - 880 - 890 - 900 - 910
920 - 930 - 940 - 950 - 960 - 970 - 980 - 990 - 1000

Notes: ____

____

____

____

Reg Reps: 10 - 20 - 30 - 40 - 50 - 60 - 70 - 80 - 90 - 100 - 110 - 120 - 130
140 - 150 - 160 - 170 - 180 - 190 - 200 - 210 - 220 - 230 - 240 - 250 - 260
270 - 280 - 290 - 300 - 310 - 320 - 330 - 340 - 350 - 360 - 370 - 380 - 390
400 - 410 - 420 - 430 - 440 - 450 - 460 - 470 - 480 - 490 - 500 - 510 - 520
530 - 540 - 550 - 560 - 570 - 580 - 590 - 600 - 610 - 620 - 630 - 640 - 650
660 - 670 - 680 - 690 - 700 - 710 - 720 - 730 - 740 - 750 - 760 - 770 - 780
790 - 800 - 810 - 820 - 830 - 840 - 850 - 860 - 870 - 880 - 890 - 900 - 910
920 - 930 - 940 - 950 - 960 - 970 - 980 - 990 - 1000

Notes: ____

____

____

______________________________ **End Date:** ____________

Technique:________________ **L - R - Alt. Start Date:** _______
Notes:________________________

Slow Reps: 10 - 20 - 30 - 40 - 50 - 60 - 70 - 80 - 90 - 100 - 110 - 120 - 130
140 - 150 - 160 - 170 - 180 - 190 - 200 - 210 - 220 - 230 - 240 - 250 - 260
270 - 280 - 290 - 300 - 310 - 320 - 330 - 340 - 350 - 360 - 370 - 380 - 390
400 - 410 - 420 - 430 - 440 - 450 - 460 - 470 - 480 - 490 - 500 - 510 - 520
530 - 540 - 550 - 560 - 570 - 580 - 590 - 600 - 610 - 620 - 630 - 640 - 650
660 - 670 - 680 - 690 - 700 - 710 - 720 - 730 - 740 - 750 - 760 - 770 - 780
790 - 800 - 810 - 820 - 830 - 840 - 850 - 860 - 870 - 880 - 890 - 900 - 910
920 - 930 - 940 - 950 - 960 - 970 - 980 - 990 - 1000

Reg Reps: 10 - 20 - 30 - 40 - 50 - 60 - 70 - 80 - 90 - 100 - 110 - 120 - 130
140 - 150 - 160 - 170 - 180 - 190 - 200 - 210 - 220 - 230 - 240 - 250 - 260
270 - 280 - 290 - 300 - 310 - 320 - 330 - 340 - 350 - 360 - 370 - 380 - 390
400 - 410 - 420 - 430 - 440 - 450 - 460 - 470 - 480 - 490 - 500 - 510 - 520
530 - 540 - 550 - 560 - 570 - 580 - 590 - 600 - 610 - 620 - 630 - 640 - 650
660 - 670 - 680 - 690 - 700 - 710 - 720 - 730 - 740 - 750 - 760 - 770 - 780
790 - 800 - 810 - 820 - 830 - 840 - 850 - 860 - 870 - 880 - 890 - 900 - 910
920 - 930 - 940 - 950 - 960 - 970 - 980 - 990 - 1000
Notes: ________________________

Reg Reps: 10 - 20 - 30 - 40 - 50 - 60 - 70 - 80 - 90 - 100 - 110 - 120 - 130
140 - 150 - 160 - 170 - 180 - 190 - 200 - 210 - 220 - 230 - 240 - 250 - 260
270 - 280 - 290 - 300 - 310 - 320 - 330 - 340 - 350 - 360 - 370 - 380 - 390
400 - 410 - 420 - 430 - 440 - 450 - 460 - 470 - 480 - 490 - 500 - 510 - 520
530 - 540 - 550 - 560 - 570 - 580 - 590 - 600 - 610 - 620 - 630 - 640 - 650
660 - 670 - 680 - 690 - 700 - 710 - 720 - 730 - 740 - 750 - 760 - 770 - 780
790 - 800 - 810 - 820 - 830 - 840 - 850 - 860 - 870 - 880 - 890 - 900 - 910
920 - 930 - 940 - 950 - 960 - 970 - 980 - 990 - 1000
Notes: ________________________

Reg Reps: 10 - 20 - 30 - 40 - 50 - 60 - 70 - 80 - 90 - 100 - 110 - 120 - 130 - 140 - 150 - 160 - 170 - 180 - 190 - 200 - 210 - 220 - 230 - 240 - 250 - 260 - 270 - 280 - 290 - 300 - 310 - 320 - 330 - 340 - 350 - 360 - 370 - 380 - 390 - 400 - 410 - 420 - 430 - 440 - 450 - 460 - 470 - 480 - 490 - 500 - 510 - 520 - 530 - 540 - 550 - 560 - 570 - 580 - 590 - 600 - 610 - 620 - 630 - 640 - 650 - 660 - 670 - 680 - 690 - 700 - 710 - 720 - 730 - 740 - 750 - 760 - 770 - 780 - 790 - 800 - 810 - 820 - 830 - 840 - 850 - 860 - 870 - 880 - 890 - 900 - 910 - 920 - 930 - 940 - 950 - 960 - 970 - 980 - 990 - 1000

Notes: ______________________________

____

____

____

Reg Reps: 10 - 20 - 30 - 40 - 50 - 60 - 70 - 80 - 90 - 100 - 110 - 120 - 130 - 140 - 150 - 160 - 170 - 180 - 190 - 200 - 210 - 220 - 230 - 240 - 250 - 260 - 270 - 280 - 290 - 300 - 310 - 320 - 330 - 340 - 350 - 360 - 370 - 380 - 390 - 400 - 410 - 420 - 430 - 440 - 450 - 460 - 470 - 480 - 490 - 500 - 510 - 520 - 530 - 540 - 550 - 560 - 570 - 580 - 590 - 600 - 610 - 620 - 630 - 640 - 650 - 660 - 670 - 680 - 690 - 700 - 710 - 720 - 730 - 740 - 750 - 760 - 770 - 780 - 790 - 800 - 810 - 820 - 830 - 840 - 850 - 860 - 870 - 880 - 890 - 900 - 910 - 920 - 930 - 940 - 950 - 960 - 970 - 980 - 990 - 1000

Notes: ______________________________

____

____

____

Reg Reps: 10 - 20 - 30 - 40 - 50 - 60 - 70 - 80 - 90 - 100 - 110 - 120 - 130 - 140 - 150 - 160 - 170 - 180 - 190 - 200 - 210 - 220 - 230 - 240 - 250 - 260 - 270 - 280 - 290 - 300 - 310 - 320 - 330 - 340 - 350 - 360 - 370 - 380 - 390 - 400 - 410 - 420 - 430 - 440 - 450 - 460 - 470 - 480 - 490 - 500 - 510 - 520 - 530 - 540 - 550 - 560 - 570 - 580 - 590 - 600 - 610 - 620 - 630 - 640 - 650 - 660 - 670 - 680 - 690 - 700 - 710 - 720 - 730 - 740 - 750 - 760 - 770 - 780 - 790 - 800 - 810 - 820 - 830 - 840 - 850 - 860 - 870 - 880 - 890 - 900 - 910 - 920 - 930 - 940 - 950 - 960 - 970 - 980 - 990 - 1000

Notes: ______________________________

____

____

______________________________ **End Date:** ____________

Technique:________________________ **L - R - Alt. Start Date:** _______
Notes:____

Slow Reps: 10 - 20 - 30 - 40 - 50 - 60 - 70 - 80 - 90 - 100 - 110 - 120 - 130
140 - 150 - 160 - 170 - 180 - 190 - 200 - 210 - 220 - 230 - 240 - 250 - 260
270 - 280 - 290 - 300 - 310 - 320 - 330 - 340 - 350 - 360 - 370 - 380 - 390
400 - 410 - 420 - 430 - 440 - 450 - 460 - 470 - 480 - 490 - 500 - 510 - 520
530 - 540 - 550 - 560 - 570 - 580 - 590 - 600 - 610 - 620 - 630 - 640 - 650
660 - 670 - 680 - 690 - 700 - 710 - 720 - 730 - 740 - 750 - 760 - 770 - 780
790 - 800 - 810 - 820 - 830 - 840 - 850 - 860 - 870 - 880 - 890 - 900 - 910
920 - 930 - 940 - 950 - 960 - 970 - 980 - 990 - 1000

Reg Reps: 10 - 20 - 30 - 40 - 50 - 60 - 70 - 80 - 90 - 100 - 110 - 120 - 130
140 - 150 - 160 - 170 - 180 - 190 - 200 - 210 - 220 - 230 - 240 - 250 - 260
270 - 280 - 290 - 300 - 310 - 320 - 330 - 340 - 350 - 360 - 370 - 380 - 390
400 - 410 - 420 - 430 - 440 - 450 - 460 - 470 - 480 - 490 - 500 - 510 - 520
530 - 540 - 550 - 560 - 570 - 580 - 590 - 600 - 610 - 620 - 630 - 640 - 650
660 - 670 - 680 - 690 - 700 - 710 - 720 - 730 - 740 - 750 - 760 - 770 - 780
790 - 800 - 810 - 820 - 830 - 840 - 850 - 860 - 870 - 880 - 890 - 900 - 910
920 - 930 - 940 - 950 - 960 - 970 - 980 - 990 - 1000
Notes: ____

Reg Reps: 10 - 20 - 30 - 40 - 50 - 60 - 70 - 80 - 90 - 100 - 110 - 120 - 130
140 - 150 - 160 - 170 - 180 - 190 - 200 - 210 - 220 - 230 - 240 - 250 - 260
270 - 280 - 290 - 300 - 310 - 320 - 330 - 340 - 350 - 360 - 370 - 380 - 390
400 - 410 - 420 - 430 - 440 - 450 - 460 - 470 - 480 - 490 - 500 - 510 - 520
530 - 540 - 550 - 560 - 570 - 580 - 590 - 600 - 610 - 620 - 630 - 640 - 650
660 - 670 - 680 - 690 - 700 - 710 - 720 - 730 - 740 - 750 - 760 - 770 - 780
790 - 800 - 810 - 820 - 830 - 840 - 850 - 860 - 870 - 880 - 890 - 900 - 910
920 - 930 - 940 - 950 - 960 - 970 - 980 - 990 - 1000
Notes: ____

Reg Reps: 10 - 20 - 30 - 40 - 50 - 60 - 70 - 80 - 90 - 100 - 110 - 120 - 130
140 - 150 - 160 - 170 - 180 - 190 - 200 - 210 - 220 - 230 - 240 - 250 - 260
270 - 280 - 290 - 300 - 310 - 320 - 330 - 340 - 350 - 360 - 370 - 380 - 390
400 - 410 - 420 - 430 - 440 - 450 - 460 - 470 - 480 - 490 - 500 - 510 - 520
530 - 540 - 550 - 560 - 570 - 580 - 590 - 600 - 610 - 620 - 630 - 640 - 650
660 - 670 - 680 - 690 - 700 - 710 - 720 - 730 - 740 - 750 - 760 - 770 - 780
790 - 800 - 810 - 820 - 830 - 840 - 850 - 860 - 870 - 880 - 890 - 900 - 910
920 - 930 - 940 - 950 - 960 - 970 - 980 - 990 - 1000

Notes: ______________________________

____

____

____

Reg Reps: 10 - 20 - 30 - 40 - 50 - 60 - 70 - 80 - 90 - 100 - 110 - 120 - 130
140 - 150 - 160 - 170 - 180 - 190 - 200 - 210 - 220 - 230 - 240 - 250 - 260
270 - 280 - 290 - 300 - 310 - 320 - 330 - 340 - 350 - 360 - 370 - 380 - 390
400 - 410 - 420 - 430 - 440 - 450 - 460 - 470 - 480 - 490 - 500 - 510 - 520
530 - 540 - 550 - 560 - 570 - 580 - 590 - 600 - 610 - 620 - 630 - 640 - 650
660 - 670 - 680 - 690 - 700 - 710 - 720 - 730 - 740 - 750 - 760 - 770 - 780
790 - 800 - 810 - 820 - 830 - 840 - 850 - 860 - 870 - 880 - 890 - 900 - 910
920 - 930 - 940 - 950 - 960 - 970 - 980 - 990 - 1000

Notes: ______________________________

____

____

____

Reg Reps: 10 - 20 - 30 - 40 - 50 - 60 - 70 - 80 - 90 - 100 - 110 - 120 - 130
140 - 150 - 160 - 170 - 180 - 190 - 200 - 210 - 220 - 230 - 240 - 250 - 260
270 - 280 - 290 - 300 - 310 - 320 - 330 - 340 - 350 - 360 - 370 - 380 - 390
400 - 410 - 420 - 430 - 440 - 450 - 460 - 470 - 480 - 490 - 500 - 510 - 520
530 - 540 - 550 - 560 - 570 - 580 - 590 - 600 - 610 - 620 - 630 - 640 - 650
660 - 670 - 680 - 690 - 700 - 710 - 720 - 730 - 740 - 750 - 760 - 770 - 780
790 - 800 - 810 - 820 - 830 - 840 - 850 - 860 - 870 - 880 - 890 - 900 - 910
920 - 930 - 940 - 950 - 960 - 970 - 980 - 990 - 1000

Notes: ______________________________

____

____

______________________________ **End Date:** __________

Technique:________________________ **L - R - Alt. Start Date:** _______

Notes:____

Slow Reps: 10 - 20 - 30 - 40 - 50 - 60 - 70 - 80 - 90 - 100 - 110 - 120 - 130
140 - 150 - 160 - 170 - 180 - 190 - 200 - 210 - 220 - 230 - 240 - 250 - 260
270 - 280 - 290 - 300 - 310 - 320 - 330 - 340 - 350 - 360 - 370 - 380 - 390
400 - 410 - 420 - 430 - 440 - 450 - 460 - 470 - 480 - 490 - 500 - 510 - 520
530 - 540 - 550 - 560 - 570 - 580 - 590 - 600 - 610 - 620 - 630 - 640 - 650
660 - 670 - 680 - 690 - 700 - 710 - 720 - 730 - 740 - 750 - 760 - 770 - 780
790 - 800 - 810 - 820 - 830 - 840 - 850 - 860 - 870 - 880 - 890 - 900 - 910
920 - 930 - 940 - 950 - 960 - 970 - 980 - 990 - 1000

Reg Reps: 10 - 20 - 30 - 40 - 50 - 60 - 70 - 80 - 90 - 100 - 110 - 120 - 130
140 - 150 - 160 - 170 - 180 - 190 - 200 - 210 - 220 - 230 - 240 - 250 - 260
270 - 280 - 290 - 300 - 310 - 320 - 330 - 340 - 350 - 360 - 370 - 380 - 390
400 - 410 - 420 - 430 - 440 - 450 - 460 - 470 - 480 - 490 - 500 - 510 - 520
530 - 540 - 550 - 560 - 570 - 580 - 590 - 600 - 610 - 620 - 630 - 640 - 650
660 - 670 - 680 - 690 - 700 - 710 - 720 - 730 - 740 - 750 - 760 - 770 - 780
790 - 800 - 810 - 820 - 830 - 840 - 850 - 860 - 870 - 880 - 890 - 900 - 910
920 - 930 - 940 - 950 - 960 - 970 - 980 - 990 - 1000

Notes: ____

Reg Reps: 10 - 20 - 30 - 40 - 50 - 60 - 70 - 80 - 90 - 100 - 110 - 120 - 130
140 - 150 - 160 - 170 - 180 - 190 - 200 - 210 - 220 - 230 - 240 - 250 - 260
270 - 280 - 290 - 300 - 310 - 320 - 330 - 340 - 350 - 360 - 370 - 380 - 390
400 - 410 - 420 - 430 - 440 - 450 - 460 - 470 - 480 - 490 - 500 - 510 - 520
530 - 540 - 550 - 560 - 570 - 580 - 590 - 600 - 610 - 620 - 630 - 640 - 650
660 - 670 - 680 - 690 - 700 - 710 - 720 - 730 - 740 - 750 - 760 - 770 - 780
790 - 800 - 810 - 820 - 830 - 840 - 850 - 860 - 870 - 880 - 890 - 900 - 910
920 - 930 - 940 - 950 - 960 - 970 - 980 - 990 - 1000

Notes: ____

Reg Reps: 10 - 20 - 30 - 40 - 50 - 60 - 70 - 80 - 90 - 100 - 110 - 120 - 130
140 - 150 - 160 - 170 - 180 - 190 - 200 - 210 - 220 - 230 - 240 - 250 - 260
270 - 280 - 290 - 300 - 310 - 320 - 330 - 340 - 350 - 360 - 370 - 380 - 390
400 - 410 - 420 - 430 - 440 - 450 - 460 - 470 - 480 - 490 - 500 - 510 - 520
530 - 540 - 550 - 560 - 570 - 580 - 590 - 600 - 610 - 620 - 630 - 640 - 650
660 - 670 - 680 - 690 - 700 - 710 - 720 - 730 - 740 - 750 - 760 - 770 - 780
790 - 800 - 810 - 820 - 830 - 840 - 850 - 860 - 870 - 880 - 890 - 900 - 910
920 - 930 - 940 - 950 - 960 - 970 - 980 - 990 - 1000

Notes: ______________________________

____

____

____

Reg Reps: 10 - 20 - 30 - 40 - 50 - 60 - 70 - 80 - 90 - 100 - 110 - 120 - 130
140 - 150 - 160 - 170 - 180 - 190 - 200 - 210 - 220 - 230 - 240 - 250 - 260
270 - 280 - 290 - 300 - 310 - 320 - 330 - 340 - 350 - 360 - 370 - 380 - 390
400 - 410 - 420 - 430 - 440 - 450 - 460 - 470 - 480 - 490 - 500 - 510 - 520
530 - 540 - 550 - 560 - 570 - 580 - 590 - 600 - 610 - 620 - 630 - 640 - 650
660 - 670 - 680 - 690 - 700 - 710 - 720 - 730 - 740 - 750 - 760 - 770 - 780
790 - 800 - 810 - 820 - 830 - 840 - 850 - 860 - 870 - 880 - 890 - 900 - 910
920 - 930 - 940 - 950 - 960 - 970 - 980 - 990 - 1000

Notes: ______________________________

____

____

____

Reg Reps: 10 - 20 - 30 - 40 - 50 - 60 - 70 - 80 - 90 - 100 - 110 - 120 - 130
140 - 150 - 160 - 170 - 180 - 190 - 200 - 210 - 220 - 230 - 240 - 250 - 260
270 - 280 - 290 - 300 - 310 - 320 - 330 - 340 - 350 - 360 - 370 - 380 - 390
400 - 410 - 420 - 430 - 440 - 450 - 460 - 470 - 480 - 490 - 500 - 510 - 520
530 - 540 - 550 - 560 - 570 - 580 - 590 - 600 - 610 - 620 - 630 - 640 - 650
660 - 670 - 680 - 690 - 700 - 710 - 720 - 730 - 740 - 750 - 760 - 770 - 780
790 - 800 - 810 - 820 - 830 - 840 - 850 - 860 - 870 - 880 - 890 - 900 - 910
920 - 930 - 940 - 950 - 960 - 970 - 980 - 990 - 1000

Notes: ______________________________

____

____

______________________________**End Date:**______________

Technique:________________________ **L - R - Alt. Start Date:** _______
Notes:____

Slow Reps: 10 - 20 - 30 - 40 - 50 - 60 - 70 - 80 - 90 - 100 - 110 - 120 - 130
140 - 150 - 160 - 170 - 180 - 190 - 200 - 210 - 220 - 230 - 240 - 250 - 260
270 - 280 - 290 - 300 - 310 - 320 - 330 - 340 - 350 - 360 - 370 - 380 - 390
400 - 410 - 420 - 430 - 440 - 450 - 460 - 470 - 480 - 490 - 500 - 510 - 520
530 - 540 - 550 - 560 - 570 - 580 - 590 - 600 - 610 - 620 - 630 - 640 - 650
660 - 670 - 680 - 690 - 700 - 710 - 720 - 730 - 740 - 750 - 760 - 770 - 780
790 - 800 - 810 - 820 - 830 - 840 - 850 - 860 - 870 - 880 - 890 - 900 - 910
920 - 930 - 940 - 950 - 960 - 970 - 980 - 990 - 1000

Reg Reps: 10 - 20 - 30 - 40 - 50 - 60 - 70 - 80 - 90 - 100 - 110 - 120 - 130
140 - 150 - 160 - 170 - 180 - 190 - 200 - 210 - 220 - 230 - 240 - 250 - 260
270 - 280 - 290 - 300 - 310 - 320 - 330 - 340 - 350 - 360 - 370 - 380 - 390
400 - 410 - 420 - 430 - 440 - 450 - 460 - 470 - 480 - 490 - 500 - 510 - 520
530 - 540 - 550 - 560 - 570 - 580 - 590 - 600 - 610 - 620 - 630 - 640 - 650
660 - 670 - 680 - 690 - 700 - 710 - 720 - 730 - 740 - 750 - 760 - 770 - 780
790 - 800 - 810 - 820 - 830 - 840 - 850 - 860 - 870 - 880 - 890 - 900 - 910
920 - 930 - 940 - 950 - 960 - 970 - 980 - 990 - 1000
Notes: ____

Reg Reps: 10 - 20 - 30 - 40 - 50 - 60 - 70 - 80 - 90 - 100 - 110 - 120 - 130
140 - 150 - 160 - 170 - 180 - 190 - 200 - 210 - 220 - 230 - 240 - 250 - 260
270 - 280 - 290 - 300 - 310 - 320 - 330 - 340 - 350 - 360 - 370 - 380 - 390
400 - 410 - 420 - 430 - 440 - 450 - 460 - 470 - 480 - 490 - 500 - 510 - 520
530 - 540 - 550 - 560 - 570 - 580 - 590 - 600 - 610 - 620 - 630 - 640 - 650
660 - 670 - 680 - 690 - 700 - 710 - 720 - 730 - 740 - 750 - 760 - 770 - 780
790 - 800 - 810 - 820 - 830 - 840 - 850 - 860 - 870 - 880 - 890 - 900 - 910
920 - 930 - 940 - 950 - 960 - 970 - 980 - 990 - 1000
Notes: ____

Reg Reps: 10 - 20 - 30 - 40 - 50 - 60 - 70 - 80 - 90 - 100 - 110 - 120 - 130 140 - 150 - 160 - 170 - 180 - 190 - 200 - 210 - 220 - 230 - 240 - 250 - 260 270 - 280 - 290 - 300 - 310 - 320 - 330 - 340 - 350 - 360 - 370 - 380 - 390 400 - 410 - 420 - 430 - 440 - 450 - 460 - 470 - 480 - 490 - 500 - 510 - 520 530 - 540 - 550 - 560 - 570 - 580 - 590 - 600 - 610 - 620 - 630 - 640 - 650 660 - 670 - 680 - 690 - 700 - 710 - 720 - 730 - 740 - 750 - 760 - 770 - 780 790 - 800 - 810 - 820 - 830 - 840 - 850 - 860 - 870 - 880 - 890 - 900 - 910 920 - 930 - 940 - 950 - 960 - 970 - 980 - 990 - 1000

Notes: ______________________________

____

____

____

Reg Reps: 10 - 20 - 30 - 40 - 50 - 60 - 70 - 80 - 90 - 100 - 110 - 120 - 130 140 - 150 - 160 - 170 - 180 - 190 - 200 - 210 - 220 - 230 - 240 - 250 - 260 270 - 280 - 290 - 300 - 310 - 320 - 330 - 340 - 350 - 360 - 370 - 380 - 390 400 - 410 - 420 - 430 - 440 - 450 - 460 - 470 - 480 - 490 - 500 - 510 - 520 530 - 540 - 550 - 560 - 570 - 580 - 590 - 600 - 610 - 620 - 630 - 640 - 650 660 - 670 - 680 - 690 - 700 - 710 - 720 - 730 - 740 - 750 - 760 - 770 - 780 790 - 800 - 810 - 820 - 830 - 840 - 850 - 860 - 870 - 880 - 890 - 900 - 910 920 - 930 - 940 - 950 - 960 - 970 - 980 - 990 - 1000

Notes: ______________________________

____

____

____

Reg Reps: 10 - 20 - 30 - 40 - 50 - 60 - 70 - 80 - 90 - 100 - 110 - 120 - 130 140 - 150 - 160 - 170 - 180 - 190 - 200 - 210 - 220 - 230 - 240 - 250 - 260 270 - 280 - 290 - 300 - 310 - 320 - 330 - 340 - 350 - 360 - 370 - 380 - 390 400 - 410 - 420 - 430 - 440 - 450 - 460 - 470 - 480 - 490 - 500 - 510 - 520 530 - 540 - 550 - 560 - 570 - 580 - 590 - 600 - 610 - 620 - 630 - 640 - 650 660 - 670 - 680 - 690 - 700 - 710 - 720 - 730 - 740 - 750 - 760 - 770 - 780 790 - 800 - 810 - 820 - 830 - 840 - 850 - 860 - 870 - 880 - 890 - 900 - 910 920 - 930 - 940 - 950 - 960 - 970 - 980 - 990 - 1000

Notes: ______________________________

____

____

______________________________ **End Date:** ____________

Technique:______________________ **L - R - Alt. Start Date:** _______

Notes:____

Slow Reps: 10 - 20 - 30 - 40 - 50 - 60 - 70 - 80 - 90 - 100 - 110 - 120 - 130
140 - 150 - 160 - 170 - 180 - 190 - 200 - 210 - 220 - 230 - 240 - 250 - 260
270 - 280 - 290 - 300 - 310 - 320 - 330 - 340 - 350 - 360 - 370 - 380 - 390
400 - 410 - 420 - 430 - 440 - 450 - 460 - 470 - 480 - 490 - 500 - 510 - 520
530 - 540 - 550 - 560 - 570 - 580 - 590 - 600 - 610 - 620 - 630 - 640 - 650
660 - 670 - 680 - 690 - 700 - 710 - 720 - 730 - 740 - 750 - 760 - 770 - 780
790 - 800 - 810 - 820 - 830 - 840 - 850 - 860 - 870 - 880 - 890 - 900 - 910
920 - 930 - 940 - 950 - 960 - 970 - 980 - 990 - 1000

Reg Reps: 10 - 20 - 30 - 40 - 50 - 60 - 70 - 80 - 90 - 100 - 110 - 120 - 130
140 - 150 - 160 - 170 - 180 - 190 - 200 - 210 - 220 - 230 - 240 - 250 - 260
270 - 280 - 290 - 300 - 310 - 320 - 330 - 340 - 350 - 360 - 370 - 380 - 390
400 - 410 - 420 - 430 - 440 - 450 - 460 - 470 - 480 - 490 - 500 - 510 - 520
530 - 540 - 550 - 560 - 570 - 580 - 590 - 600 - 610 - 620 - 630 - 640 - 650
660 - 670 - 680 - 690 - 700 - 710 - 720 - 730 - 740 - 750 - 760 - 770 - 780
790 - 800 - 810 - 820 - 830 - 840 - 850 - 860 - 870 - 880 - 890 - 900 - 910
920 - 930 - 940 - 950 - 960 - 970 - 980 - 990 - 1000

Notes: ____

Reg Reps: 10 - 20 - 30 - 40 - 50 - 60 - 70 - 80 - 90 - 100 - 110 - 120 - 130
140 - 150 - 160 - 170 - 180 - 190 - 200 - 210 - 220 - 230 - 240 - 250 - 260
270 - 280 - 290 - 300 - 310 - 320 - 330 - 340 - 350 - 360 - 370 - 380 - 390
400 - 410 - 420 - 430 - 440 - 450 - 460 - 470 - 480 - 490 - 500 - 510 - 520
530 - 540 - 550 - 560 - 570 - 580 - 590 - 600 - 610 - 620 - 630 - 640 - 650
660 - 670 - 680 - 690 - 700 - 710 - 720 - 730 - 740 - 750 - 760 - 770 - 780
790 - 800 - 810 - 820 - 830 - 840 - 850 - 860 - 870 - 880 - 890 - 900 - 910
920 - 930 - 940 - 950 - 960 - 970 - 980 - 990 - 1000

Notes: ____

Reg Reps: 10 - 20 - 30 - 40 - 50 - 60 - 70 - 80 - 90 - 100 - 110 - 120 - 130 140 - 150 - 160 - 170 - 180 - 190 - 200 - 210 - 220 - 230 - 240 - 250 - 260 270 - 280 - 290 - 300 - 310 - 320 - 330 - 340 - 350 - 360 - 370 - 380 - 390 400 - 410 - 420 - 430 - 440 - 450 - 460 - 470 - 480 - 490 - 500 - 510 - 520 530 - 540 - 550 - 560 - 570 - 580 - 590 - 600 - 610 - 620 - 630 - 640 - 650 660 - 670 - 680 - 690 - 700 - 710 - 720 - 730 - 740 - 750 - 760 - 770 - 780 790 - 800 - 810 - 820 - 830 - 840 - 850 - 860 - 870 - 880 - 890 - 900 - 910 920 - 930 - 940 - 950 - 960 - 970 - 980 - 990 - 1000

Notes: ______________________________

____

____

____

Reg Reps: 10 - 20 - 30 - 40 - 50 - 60 - 70 - 80 - 90 - 100 - 110 - 120 - 130 140 - 150 - 160 - 170 - 180 - 190 - 200 - 210 - 220 - 230 - 240 - 250 - 260 270 - 280 - 290 - 300 - 310 - 320 - 330 - 340 - 350 - 360 - 370 - 380 - 390 400 - 410 - 420 - 430 - 440 - 450 - 460 - 470 - 480 - 490 - 500 - 510 - 520 530 - 540 - 550 - 560 - 570 - 580 - 590 - 600 - 610 - 620 - 630 - 640 - 650 660 - 670 - 680 - 690 - 700 - 710 - 720 - 730 - 740 - 750 - 760 - 770 - 780 790 - 800 - 810 - 820 - 830 - 840 - 850 - 860 - 870 - 880 - 890 - 900 - 910 920 - 930 - 940 - 950 - 960 - 970 - 980 - 990 - 1000

Notes: ______________________________

____

____

____

Reg Reps: 10 - 20 - 30 - 40 - 50 - 60 - 70 - 80 - 90 - 100 - 110 - 120 - 130 140 - 150 - 160 - 170 - 180 - 190 - 200 - 210 - 220 - 230 - 240 - 250 - 260 270 - 280 - 290 - 300 - 310 - 320 - 330 - 340 - 350 - 360 - 370 - 380 - 390 400 - 410 - 420 - 430 - 440 - 450 - 460 - 470 - 480 - 490 - 500 - 510 - 520 530 - 540 - 550 - 560 - 570 - 580 - 590 - 600 - 610 - 620 - 630 - 640 - 650 660 - 670 - 680 - 690 - 700 - 710 - 720 - 730 - 740 - 750 - 760 - 770 - 780 790 - 800 - 810 - 820 - 830 - 840 - 850 - 860 - 870 - 880 - 890 - 900 - 910 920 - 930 - 940 - 950 - 960 - 970 - 980 - 990 - 1000

Notes: ______________________________

____

____

______________________________ **End Date:** ______________

Technique:_____________________ **L - R - Alt. Start Date:** ______

Notes:____

Slow Reps: 10 - 20 - 30 - 40 - 50 - 60 - 70 - 80 - 90 - 100 - 110 - 120 - 130
140 - 150 - 160 - 170 - 180 - 190 - 200 - 210 - 220 - 230 - 240 - 250 - 260
270 - 280 - 290 - 300 - 310 - 320 - 330 - 340 - 350 - 360 - 370 - 380 - 390
400 - 410 - 420 - 430 - 440 - 450 - 460 - 470 - 480 - 490 - 500 - 510 - 520
530 - 540 - 550 - 560 - 570 - 580 - 590 - 600 - 610 - 620 - 630 - 640 - 650
660 - 670 - 680 - 690 - 700 - 710 - 720 - 730 - 740 - 750 - 760 - 770 - 780
790 - 800 - 810 - 820 - 830 - 840 - 850 - 860 - 870 - 880 - 890 - 900 - 910
920 - 930 - 940 - 950 - 960 - 970 - 980 - 990 - 1000

Reg Reps: 10 - 20 - 30 - 40 - 50 - 60 - 70 - 80 - 90 - 100 - 110 - 120 - 130
140 - 150 - 160 - 170 - 180 - 190 - 200 - 210 - 220 - 230 - 240 - 250 - 260
270 - 280 - 290 - 300 - 310 - 320 - 330 - 340 - 350 - 360 - 370 - 380 - 390
400 - 410 - 420 - 430 - 440 - 450 - 460 - 470 - 480 - 490 - 500 - 510 - 520
530 - 540 - 550 - 560 - 570 - 580 - 590 - 600 - 610 - 620 - 630 - 640 - 650
660 - 670 - 680 - 690 - 700 - 710 - 720 - 730 - 740 - 750 - 760 - 770 - 780
790 - 800 - 810 - 820 - 830 - 840 - 850 - 860 - 870 - 880 - 890 - 900 - 910
920 - 930 - 940 - 950 - 960 - 970 - 980 - 990 - 1000

Notes: ____

Reg Reps: 10 - 20 - 30 - 40 - 50 - 60 - 70 - 80 - 90 - 100 - 110 - 120 - 130
140 - 150 - 160 - 170 - 180 - 190 - 200 - 210 - 220 - 230 - 240 - 250 - 260
270 - 280 - 290 - 300 - 310 - 320 - 330 - 340 - 350 - 360 - 370 - 380 - 390
400 - 410 - 420 - 430 - 440 - 450 - 460 - 470 - 480 - 490 - 500 - 510 - 520
530 - 540 - 550 - 560 - 570 - 580 - 590 - 600 - 610 - 620 - 630 - 640 - 650
660 - 670 - 680 - 690 - 700 - 710 - 720 - 730 - 740 - 750 - 760 - 770 - 780
790 - 800 - 810 - 820 - 830 - 840 - 850 - 860 - 870 - 880 - 890 - 900 - 910
920 - 930 - 940 - 950 - 960 - 970 - 980 - 990 - 1000

Notes: ____

Reg Reps: 10 - 20 - 30 - 40 - 50 - 60 - 70 - 80 - 90 - 100 - 110 - 120 - 130
140 - 150 - 160 - 170 - 180 - 190 - 200 - 210 - 220 - 230 - 240 - 250 - 260
270 - 280 - 290 - 300 - 310 - 320 - 330 - 340 - 350 - 360 - 370 - 380 - 390
400 - 410 - 420 - 430 - 440 - 450 - 460 - 470 - 480 - 490 - 500 - 510 - 520
530 - 540 - 550 - 560 - 570 - 580 - 590 - 600 - 610 - 620 - 630 - 640 - 650
660 - 670 - 680 - 690 - 700 - 710 - 720 - 730 - 740 - 750 - 760 - 770 - 780
790 - 800 - 810 - 820 - 830 - 840 - 850 - 860 - 870 - 880 - 890 - 900 - 910
920 - 930 - 940 - 950 - 960 - 970 - 980 - 990 - 1000

Notes: ____

____

____

____

Reg Reps: 10 - 20 - 30 - 40 - 50 - 60 - 70 - 80 - 90 - 100 - 110 - 120 - 130
140 - 150 - 160 - 170 - 180 - 190 - 200 - 210 - 220 - 230 - 240 - 250 - 260
270 - 280 - 290 - 300 - 310 - 320 - 330 - 340 - 350 - 360 - 370 - 380 - 390
400 - 410 - 420 - 430 - 440 - 450 - 460 - 470 - 480 - 490 - 500 - 510 - 520
530 - 540 - 550 - 560 - 570 - 580 - 590 - 600 - 610 - 620 - 630 - 640 - 650
660 - 670 - 680 - 690 - 700 - 710 - 720 - 730 - 740 - 750 - 760 - 770 - 780
790 - 800 - 810 - 820 - 830 - 840 - 850 - 860 - 870 - 880 - 890 - 900 - 910
920 - 930 - 940 - 950 - 960 - 970 - 980 - 990 - 1000

Notes: ____

____

____

____

Reg Reps: 10 - 20 - 30 - 40 - 50 - 60 - 70 - 80 - 90 - 100 - 110 - 120 - 130
140 - 150 - 160 - 170 - 180 - 190 - 200 - 210 - 220 - 230 - 240 - 250 - 260
270 - 280 - 290 - 300 - 310 - 320 - 330 - 340 - 350 - 360 - 370 - 380 - 390
400 - 410 - 420 - 430 - 440 - 450 - 460 - 470 - 480 - 490 - 500 - 510 - 520
530 - 540 - 550 - 560 - 570 - 580 - 590 - 600 - 610 - 620 - 630 - 640 - 650
660 - 670 - 680 - 690 - 700 - 710 - 720 - 730 - 740 - 750 - 760 - 770 - 780
790 - 800 - 810 - 820 - 830 - 840 - 850 - 860 - 870 - 880 - 890 - 900 - 910
920 - 930 - 940 - 950 - 960 - 970 - 980 - 990 - 1000

Notes: ____

____

____

______________________________ **End Date:** ____________

Technique:____________________ **L - R - Alt. Start Date:** _______
Notes:____

Slow Reps: 10 - 20 - 30 - 40 - 50 - 60 - 70 - 80 - 90 - 100 - 110 - 120 - 130
140 - 150 - 160 - 170 - 180 - 190 - 200 - 210 - 220 - 230 - 240 - 250 - 260
270 - 280 - 290 - 300 - 310 - 320 - 330 - 340 - 350 - 360 - 370 - 380 - 390
400 - 410 - 420 - 430 - 440 - 450 - 460 - 470 - 480 - 490 - 500 - 510 - 520
530 - 540 - 550 - 560 - 570 - 580 - 590 - 600 - 610 - 620 - 630 - 640 - 650
660 - 670 - 680 - 690 - 700 - 710 - 720 - 730 - 740 - 750 - 760 - 770 - 780
790 - 800 - 810 - 820 - 830 - 840 - 850 - 860 - 870 - 880 - 890 - 900 - 910
920 - 930 - 940 - 950 - 960 - 970 - 980 - 990 - 1000

Reg Reps: 10 - 20 - 30 - 40 - 50 - 60 - 70 - 80 - 90 - 100 - 110 - 120 - 130
140 - 150 - 160 - 170 - 180 - 190 - 200 - 210 - 220 - 230 - 240 - 250 - 260
270 - 280 - 290 - 300 - 310 - 320 - 330 - 340 - 350 - 360 - 370 - 380 - 390
400 - 410 - 420 - 430 - 440 - 450 - 460 - 470 - 480 - 490 - 500 - 510 - 520
530 - 540 - 550 - 560 - 570 - 580 - 590 - 600 - 610 - 620 - 630 - 640 - 650
660 - 670 - 680 - 690 - 700 - 710 - 720 - 730 - 740 - 750 - 760 - 770 - 780
790 - 800 - 810 - 820 - 830 - 840 - 850 - 860 - 870 - 880 - 890 - 900 - 910
920 - 930 - 940 - 950 - 960 - 970 - 980 - 990 - 1000
Notes: ____

Reg Reps: 10 - 20 - 30 - 40 - 50 - 60 - 70 - 80 - 90 - 100 - 110 - 120 - 130
140 - 150 - 160 - 170 - 180 - 190 - 200 - 210 - 220 - 230 - 240 - 250 - 260
270 - 280 - 290 - 300 - 310 - 320 - 330 - 340 - 350 - 360 - 370 - 380 - 390
400 - 410 - 420 - 430 - 440 - 450 - 460 - 470 - 480 - 490 - 500 - 510 - 520
530 - 540 - 550 - 560 - 570 - 580 - 590 - 600 - 610 - 620 - 630 - 640 - 650
660 - 670 - 680 - 690 - 700 - 710 - 720 - 730 - 740 - 750 - 760 - 770 - 780
790 - 800 - 810 - 820 - 830 - 840 - 850 - 860 - 870 - 880 - 890 - 900 - 910
920 - 930 - 940 - 950 - 960 - 970 - 980 - 990 - 1000
Notes: ____

Reg Reps: 10 - 20 - 30 - 40 - 50 - 60 - 70 - 80 - 90 - 100 - 110 - 120 - 130
140 - 150 - 160 - 170 - 180 - 190 - 200 - 210 - 220 - 230 - 240 - 250 - 260
270 - 280 - 290 - 300 - 310 - 320 - 330 - 340 - 350 - 360 - 370 - 380 - 390
400 - 410 - 420 - 430 - 440 - 450 - 460 - 470 - 480 - 490 - 500 - 510 - 520
530 - 540 - 550 - 560 - 570 - 580 - 590 - 600 - 610 - 620 - 630 - 640 - 650
660 - 670 - 680 - 690 - 700 - 710 - 720 - 730 - 740 - 750 - 760 - 770 - 780
790 - 800 - 810 - 820 - 830 - 840 - 850 - 860 - 870 - 880 - 890 - 900 - 910
920 - 930 - 940 - 950 - 960 - 970 - 980 - 990 - 1000

Notes: ____________________

Reg Reps: 10 - 20 - 30 - 40 - 50 - 60 - 70 - 80 - 90 - 100 - 110 - 120 - 130
140 - 150 - 160 - 170 - 180 - 190 - 200 - 210 - 220 - 230 - 240 - 250 - 260
270 - 280 - 290 - 300 - 310 - 320 - 330 - 340 - 350 - 360 - 370 - 380 - 390
400 - 410 - 420 - 430 - 440 - 450 - 460 - 470 - 480 - 490 - 500 - 510 - 520
530 - 540 - 550 - 560 - 570 - 580 - 590 - 600 - 610 - 620 - 630 - 640 - 650
660 - 670 - 680 - 690 - 700 - 710 - 720 - 730 - 740 - 750 - 760 - 770 - 780
790 - 800 - 810 - 820 - 830 - 840 - 850 - 860 - 870 - 880 - 890 - 900 - 910
920 - 930 - 940 - 950 - 960 - 970 - 980 - 990 - 1000

Notes: ____________________

Reg Reps: 10 - 20 - 30 - 40 - 50 - 60 - 70 - 80 - 90 - 100 - 110 - 120 - 130
140 - 150 - 160 - 170 - 180 - 190 - 200 - 210 - 220 - 230 - 240 - 250 - 260
270 - 280 - 290 - 300 - 310 - 320 - 330 - 340 - 350 - 360 - 370 - 380 - 390
400 - 410 - 420 - 430 - 440 - 450 - 460 - 470 - 480 - 490 - 500 - 510 - 520
530 - 540 - 550 - 560 - 570 - 580 - 590 - 600 - 610 - 620 - 630 - 640 - 650
660 - 670 - 680 - 690 - 700 - 710 - 720 - 730 - 740 - 750 - 760 - 770 - 780
790 - 800 - 810 - 820 - 830 - 840 - 850 - 860 - 870 - 880 - 890 - 900 - 910
920 - 930 - 940 - 950 - 960 - 970 - 980 - 990 - 1000

Notes: ____________________

End Date: ____________

Technique:______________________ **L - R - Alt. Start Date:** _______

Notes:____

____

____

____

Slow Reps: 10 - 20 - 30 - 40 - 50 - 60 - 70 - 80 - 90 - 100 - 110 - 120 - 130
140 - 150 - 160 - 170 - 180 - 190 - 200 - 210 - 220 - 230 - 240 - 250 - 260
270 - 280 - 290 - 300 - 310 - 320 - 330 - 340 - 350 - 360 - 370 - 380 - 390
400 - 410 - 420 - 430 - 440 - 450 - 460 - 470 - 480 - 490 - 500 - 510 - 520
530 - 540 - 550 - 560 - 570 - 580 - 590 - 600 - 610 - 620 - 630 - 640 - 650
660 - 670 - 680 - 690 - 700 - 710 - 720 - 730 - 740 - 750 - 760 - 770 - 780
790 - 800 - 810 - 820 - 830 - 840 - 850 - 860 - 870 - 880 - 890 - 900 - 910
920 - 930 - 940 - 950 - 960 - 970 - 980 - 990 - 1000

Reg Reps: 10 - 20 - 30 - 40 - 50 - 60 - 70 - 80 - 90 - 100 - 110 - 120 - 130
140 - 150 - 160 - 170 - 180 - 190 - 200 - 210 - 220 - 230 - 240 - 250 - 260
270 - 280 - 290 - 300 - 310 - 320 - 330 - 340 - 350 - 360 - 370 - 380 - 390
400 - 410 - 420 - 430 - 440 - 450 - 460 - 470 - 480 - 490 - 500 - 510 - 520
530 - 540 - 550 - 560 - 570 - 580 - 590 - 600 - 610 - 620 - 630 - 640 - 650
660 - 670 - 680 - 690 - 700 - 710 - 720 - 730 - 740 - 750 - 760 - 770 - 780
790 - 800 - 810 - 820 - 830 - 840 - 850 - 860 - 870 - 880 - 890 - 900 - 910
920 - 930 - 940 - 950 - 960 - 970 - 980 - 990 - 1000

Notes: ____

____

____

____

Reg Reps: 10 - 20 - 30 - 40 - 50 - 60 - 70 - 80 - 90 - 100 - 110 - 120 - 130
140 - 150 - 160 - 170 - 180 - 190 - 200 - 210 - 220 - 230 - 240 - 250 - 260
270 - 280 - 290 - 300 - 310 - 320 - 330 - 340 - 350 - 360 - 370 - 380 - 390
400 - 410 - 420 - 430 - 440 - 450 - 460 - 470 - 480 - 490 - 500 - 510 - 520
530 - 540 - 550 - 560 - 570 - 580 - 590 - 600 - 610 - 620 - 630 - 640 - 650
660 - 670 - 680 - 690 - 700 - 710 - 720 - 730 - 740 - 750 - 760 - 770 - 780
790 - 800 - 810 - 820 - 830 - 840 - 850 - 860 - 870 - 880 - 890 - 900 - 910
920 - 930 - 940 - 950 - 960 - 970 - 980 - 990 - 1000

Notes: ____

____

____

____

Reg Reps: 10 - 20 - 30 - 40 - 50 - 60 - 70 - 80 - 90 - 100 - 110 - 120 - 130
140 - 150 - 160 - 170 - 180 - 190 - 200 - 210 - 220 - 230 - 240 - 250 - 260
270 - 280 - 290 - 300 - 310 - 320 - 330 - 340 - 350 - 360 - 370 - 380 - 390
400 - 410 - 420 - 430 - 440 - 450 - 460 - 470 - 480 - 490 - 500 - 510 - 520
530 - 540 - 550 - 560 - 570 - 580 - 590 - 600 - 610 - 620 - 630 - 640 - 650
660 - 670 - 680 - 690 - 700 - 710 - 720 - 730 - 740 - 750 - 760 - 770 - 780
790 - 800 - 810 - 820 - 830 - 840 - 850 - 860 - 870 - 880 - 890 - 900 - 910
920 - 930 - 940 - 950 - 960 - 970 - 980 - 990 - 1000

Notes: ______________________________

____

____

____

Reg Reps: 10 - 20 - 30 - 40 - 50 - 60 - 70 - 80 - 90 - 100 - 110 - 120 - 130
140 - 150 - 160 - 170 - 180 - 190 - 200 - 210 - 220 - 230 - 240 - 250 - 260
270 - 280 - 290 - 300 - 310 - 320 - 330 - 340 - 350 - 360 - 370 - 380 - 390
400 - 410 - 420 - 430 - 440 - 450 - 460 - 470 - 480 - 490 - 500 - 510 - 520
530 - 540 - 550 - 560 - 570 - 580 - 590 - 600 - 610 - 620 - 630 - 640 - 650
660 - 670 - 680 - 690 - 700 - 710 - 720 - 730 - 740 - 750 - 760 - 770 - 780
790 - 800 - 810 - 820 - 830 - 840 - 850 - 860 - 870 - 880 - 890 - 900 - 910
920 - 930 - 940 - 950 - 960 - 970 - 980 - 990 - 1000

Notes: ______________________________

____

____

____

Reg Reps: 10 - 20 - 30 - 40 - 50 - 60 - 70 - 80 - 90 - 100 - 110 - 120 - 130
140 - 150 - 160 - 170 - 180 - 190 - 200 - 210 - 220 - 230 - 240 - 250 - 260
270 - 280 - 290 - 300 - 310 - 320 - 330 - 340 - 350 - 360 - 370 - 380 - 390
400 - 410 - 420 - 430 - 440 - 450 - 460 - 470 - 480 - 490 - 500 - 510 - 520
530 - 540 - 550 - 560 - 570 - 580 - 590 - 600 - 610 - 620 - 630 - 640 - 650
660 - 670 - 680 - 690 - 700 - 710 - 720 - 730 - 740 - 750 - 760 - 770 - 780
790 - 800 - 810 - 820 - 830 - 840 - 850 - 860 - 870 - 880 - 890 - 900 - 910
920 - 930 - 940 - 950 - 960 - 970 - 980 - 990 - 1000

Notes: ______________________________

____

____

______________________________ **End Date:** ____________

Technique:____________________ **L - R - Alt. Start Date:** ________

Notes:____

____

____

____

Slow Reps: 10 - 20 - 30 - 40 - 50 - 60 - 70 - 80 - 90 - 100 - 110 - 120 - 130
140 - 150 - 160 - 170 - 180 - 190 - 200 - 210 - 220 - 230 - 240 - 250 - 260
270 - 280 - 290 - 300 - 310 - 320 - 330 - 340 - 350 - 360 - 370 - 380 - 390
400 - 410 - 420 - 430 - 440 - 450 - 460 - 470 - 480 - 490 - 500 - 510 - 520
530 - 540 - 550 - 560 - 570 - 580 - 590 - 600 - 610 - 620 - 630 - 640 - 650
660 - 670 - 680 - 690 - 700 - 710 - 720 - 730 - 740 - 750 - 760 - 770 - 780
790 - 800 - 810 - 820 - 830 - 840 - 850 - 860 - 870 - 880 - 890 - 900 - 910
920 - 930 - 940 - 950 - 960 - 970 - 980 - 990 - 1000

Reg Reps: 10 - 20 - 30 - 40 - 50 - 60 - 70 - 80 - 90 - 100 - 110 - 120 - 130
140 - 150 - 160 - 170 - 180 - 190 - 200 - 210 - 220 - 230 - 240 - 250 - 260
270 - 280 - 290 - 300 - 310 - 320 - 330 - 340 - 350 - 360 - 370 - 380 - 390
400 - 410 - 420 - 430 - 440 - 450 - 460 - 470 - 480 - 490 - 500 - 510 - 520
530 - 540 - 550 - 560 - 570 - 580 - 590 - 600 - 610 - 620 - 630 - 640 - 650
660 - 670 - 680 - 690 - 700 - 710 - 720 - 730 - 740 - 750 - 760 - 770 - 780
790 - 800 - 810 - 820 - 830 - 840 - 850 - 860 - 870 - 880 - 890 - 900 - 910
920 - 930 - 940 - 950 - 960 - 970 - 980 - 990 - 1000

Notes: _______________________________________

____

____

____

Reg Reps: 10 - 20 - 30 - 40 - 50 - 60 - 70 - 80 - 90 - 100 - 110 - 120 - 130
140 - 150 - 160 - 170 - 180 - 190 - 200 - 210 - 220 - 230 - 240 - 250 - 260
270 - 280 - 290 - 300 - 310 - 320 - 330 - 340 - 350 - 360 - 370 - 380 - 390
400 - 410 - 420 - 430 - 440 - 450 - 460 - 470 - 480 - 490 - 500 - 510 - 520
530 - 540 - 550 - 560 - 570 - 580 - 590 - 600 - 610 - 620 - 630 - 640 - 650
660 - 670 - 680 - 690 - 700 - 710 - 720 - 730 - 740 - 750 - 760 - 770 - 780
790 - 800 - 810 - 820 - 830 - 840 - 850 - 860 - 870 - 880 - 890 - 900 - 910
920 - 930 - 940 - 950 - 960 - 970 - 980 - 990 - 1000

Notes: _______________________________________

____

____

____

Reg Reps: 10 - 20 - 30 - 40 - 50 - 60 - 70 - 80 - 90 - 100 - 110 - 120 - 130
140 - 150 - 160 - 170 - 180 - 190 - 200 - 210 - 220 - 230 - 240 - 250 - 260
270 - 280 - 290 - 300 - 310 - 320 - 330 - 340 - 350 - 360 - 370 - 380 - 390
400 - 410 - 420 - 430 - 440 - 450 - 460 - 470 - 480 - 490 - 500 - 510 - 520
530 - 540 - 550 - 560 - 570 - 580 - 590 - 600 - 610 - 620 - 630 - 640 - 650
660 - 670 - 680 - 690 - 700 - 710 - 720 - 730 - 740 - 750 - 760 - 770 - 780
790 - 800 - 810 - 820 - 830 - 840 - 850 - 860 - 870 - 880 - 890 - 900 - 910
920 - 930 - 940 - 950 - 960 - 970 - 980 - 990 - 1000

Notes: ____

____

____

____

Reg Reps: 10 - 20 - 30 - 40 - 50 - 60 - 70 - 80 - 90 - 100 - 110 - 120 - 130
140 - 150 - 160 - 170 - 180 - 190 - 200 - 210 - 220 - 230 - 240 - 250 - 260
270 - 280 - 290 - 300 - 310 - 320 - 330 - 340 - 350 - 360 - 370 - 380 - 390
400 - 410 - 420 - 430 - 440 - 450 - 460 - 470 - 480 - 490 - 500 - 510 - 520
530 - 540 - 550 - 560 - 570 - 580 - 590 - 600 - 610 - 620 - 630 - 640 - 650
660 - 670 - 680 - 690 - 700 - 710 - 720 - 730 - 740 - 750 - 760 - 770 - 780
790 - 800 - 810 - 820 - 830 - 840 - 850 - 860 - 870 - 880 - 890 - 900 - 910
920 - 930 - 940 - 950 - 960 - 970 - 980 - 990 - 1000

Notes: ____

____

____

____

Reg Reps: 10 - 20 - 30 - 40 - 50 - 60 - 70 - 80 - 90 - 100 - 110 - 120 - 130
140 - 150 - 160 - 170 - 180 - 190 - 200 - 210 - 220 - 230 - 240 - 250 - 260
270 - 280 - 290 - 300 - 310 - 320 - 330 - 340 - 350 - 360 - 370 - 380 - 390
400 - 410 - 420 - 430 - 440 - 450 - 460 - 470 - 480 - 490 - 500 - 510 - 520
530 - 540 - 550 - 560 - 570 - 580 - 590 - 600 - 610 - 620 - 630 - 640 - 650
660 - 670 - 680 - 690 - 700 - 710 - 720 - 730 - 740 - 750 - 760 - 770 - 780
790 - 800 - 810 - 820 - 830 - 840 - 850 - 860 - 870 - 880 - 890 - 900 - 910
920 - 930 - 940 - 950 - 960 - 970 - 980 - 990 - 1000

Notes: ____

____

____

______________________________ **End Date:** ____________

Technique:________________________ **L - R - Alt. Start Date:** _______

Notes:____

Slow Reps: 10 - 20 - 30 - 40 - 50 - 60 - 70 - 80 - 90 - 100 - 110 - 120 - 130
140 - 150 - 160 - 170 - 180 - 190 - 200 - 210 - 220 - 230 - 240 - 250 - 260
270 - 280 - 290 - 300 - 310 - 320 - 330 - 340 - 350 - 360 - 370 - 380 - 390
400 - 410 - 420 - 430 - 440 - 450 - 460 - 470 - 480 - 490 - 500 - 510 - 520
530 - 540 - 550 - 560 - 570 - 580 - 590 - 600 - 610 - 620 - 630 - 640 - 650
660 - 670 - 680 - 690 - 700 - 710 - 720 - 730 - 740 - 750 - 760 - 770 - 780
790 - 800 - 810 - 820 - 830 - 840 - 850 - 860 - 870 - 880 - 890 - 900 - 910
920 - 930 - 940 - 950 - 960 - 970 - 980 - 990 - 1000

Reg Reps: 10 - 20 - 30 - 40 - 50 - 60 - 70 - 80 - 90 - 100 - 110 - 120 - 130
140 - 150 - 160 - 170 - 180 - 190 - 200 - 210 - 220 - 230 - 240 - 250 - 260
270 - 280 - 290 - 300 - 310 - 320 - 330 - 340 - 350 - 360 - 370 - 380 - 390
400 - 410 - 420 - 430 - 440 - 450 - 460 - 470 - 480 - 490 - 500 - 510 - 520
530 - 540 - 550 - 560 - 570 - 580 - 590 - 600 - 610 - 620 - 630 - 640 - 650
660 - 670 - 680 - 690 - 700 - 710 - 720 - 730 - 740 - 750 - 760 - 770 - 780
790 - 800 - 810 - 820 - 830 - 840 - 850 - 860 - 870 - 880 - 890 - 900 - 910
920 - 930 - 940 - 950 - 960 - 970 - 980 - 990 - 1000

Notes: ____

Reg Reps: 10 - 20 - 30 - 40 - 50 - 60 - 70 - 80 - 90 - 100 - 110 - 120 - 130
140 - 150 - 160 - 170 - 180 - 190 - 200 - 210 - 220 - 230 - 240 - 250 - 260
270 - 280 - 290 - 300 - 310 - 320 - 330 - 340 - 350 - 360 - 370 - 380 - 390
400 - 410 - 420 - 430 - 440 - 450 - 460 - 470 - 480 - 490 - 500 - 510 - 520
530 - 540 - 550 - 560 - 570 - 580 - 590 - 600 - 610 - 620 - 630 - 640 - 650
660 - 670 - 680 - 690 - 700 - 710 - 720 - 730 - 740 - 750 - 760 - 770 - 780
790 - 800 - 810 - 820 - 830 - 840 - 850 - 860 - 870 - 880 - 890 - 900 - 910
920 - 930 - 940 - 950 - 960 - 970 - 980 - 990 - 1000

Notes: ____

Reg Reps: 10 - 20 - 30 - 40 - 50 - 60 - 70 - 80 - 90 - 100 - 110 - 120 - 130
140 - 150 - 160 - 170 - 180 - 190 - 200 - 210 - 220 - 230 - 240 - 250 - 260
270 - 280 - 290 - 300 - 310 - 320 - 330 - 340 - 350 - 360 - 370 - 380 - 390
400 - 410 - 420 - 430 - 440 - 450 - 460 - 470 - 480 - 490 - 500 - 510 - 520
530 - 540 - 550 - 560 - 570 - 580 - 590 - 600 - 610 - 620 - 630 - 640 - 650
660 - 670 - 680 - 690 - 700 - 710 - 720 - 730 - 740 - 750 - 760 - 770 - 780
790 - 800 - 810 - 820 - 830 - 840 - 850 - 860 - 870 - 880 - 890 - 900 - 910
920 - 930 - 940 - 950 - 960 - 970 - 980 - 990 - 1000

Notes: ______________________________

____

____

____

Reg Reps: 10 - 20 - 30 - 40 - 50 - 60 - 70 - 80 - 90 - 100 - 110 - 120 - 130
140 - 150 - 160 - 170 - 180 - 190 - 200 - 210 - 220 - 230 - 240 - 250 - 260
270 - 280 - 290 - 300 - 310 - 320 - 330 - 340 - 350 - 360 - 370 - 380 - 390
400 - 410 - 420 - 430 - 440 - 450 - 460 - 470 - 480 - 490 - 500 - 510 - 520
530 - 540 - 550 - 560 - 570 - 580 - 590 - 600 - 610 - 620 - 630 - 640 - 650
660 - 670 - 680 - 690 - 700 - 710 - 720 - 730 - 740 - 750 - 760 - 770 - 780
790 - 800 - 810 - 820 - 830 - 840 - 850 - 860 - 870 - 880 - 890 - 900 - 910
920 - 930 - 940 - 950 - 960 - 970 - 980 - 990 - 1000

Notes: ______________________________

____

____

____

Reg Reps: 10 - 20 - 30 - 40 - 50 - 60 - 70 - 80 - 90 - 100 - 110 - 120 - 130
140 - 150 - 160 - 170 - 180 - 190 - 200 - 210 - 220 - 230 - 240 - 250 - 260
270 - 280 - 290 - 300 - 310 - 320 - 330 - 340 - 350 - 360 - 370 - 380 - 390
400 - 410 - 420 - 430 - 440 - 450 - 460 - 470 - 480 - 490 - 500 - 510 - 520
530 - 540 - 550 - 560 - 570 - 580 - 590 - 600 - 610 - 620 - 630 - 640 - 650
660 - 670 - 680 - 690 - 700 - 710 - 720 - 730 - 740 - 750 - 760 - 770 - 780
790 - 800 - 810 - 820 - 830 - 840 - 850 - 860 - 870 - 880 - 890 - 900 - 910
920 - 930 - 940 - 950 - 960 - 970 - 980 - 990 - 1000

Notes: ______________________________

____

____

______________________________ **End Date:** ____________

Technique: ______________________ **L - R - Alt. Start Date:** _______
Notes: ____

____

____

____

Slow Reps: 10 - 20 - 30 - 40 - 50 - 60 - 70 - 80 - 90 - 100 - 110 - 120 - 130
140 - 150 - 160 - 170 - 180 - 190 - 200 - 210 - 220 - 230 - 240 - 250 - 260
270 - 280 - 290 - 300 - 310 - 320 - 330 - 340 - 350 - 360 - 370 - 380 - 390
400 - 410 - 420 - 430 - 440 - 450 - 460 - 470 - 480 - 490 - 500 - 510 - 520
530 - 540 - 550 - 560 - 570 - 580 - 590 - 600 - 610 - 620 - 630 - 640 - 650
660 - 670 - 680 - 690 - 700 - 710 - 720 - 730 - 740 - 750 - 760 - 770 - 780
790 - 800 - 810 - 820 - 830 - 840 - 850 - 860 - 870 - 880 - 890 - 900 - 910
920 - 930 - 940 - 950 - 960 - 970 - 980 - 990 - 1000

Reg Reps: 10 - 20 - 30 - 40 - 50 - 60 - 70 - 80 - 90 - 100 - 110 - 120 - 130
140 - 150 - 160 - 170 - 180 - 190 - 200 - 210 - 220 - 230 - 240 - 250 - 260
270 - 280 - 290 - 300 - 310 - 320 - 330 - 340 - 350 - 360 - 370 - 380 - 390
400 - 410 - 420 - 430 - 440 - 450 - 460 - 470 - 480 - 490 - 500 - 510 - 520
530 - 540 - 550 - 560 - 570 - 580 - 590 - 600 - 610 - 620 - 630 - 640 - 650
660 - 670 - 680 - 690 - 700 - 710 - 720 - 730 - 740 - 750 - 760 - 770 - 780
790 - 800 - 810 - 820 - 830 - 840 - 850 - 860 - 870 - 880 - 890 - 900 - 910
920 - 930 - 940 - 950 - 960 - 970 - 980 - 990 - 1000
Notes: ____

____

____

____

Reg Reps: 10 - 20 - 30 - 40 - 50 - 60 - 70 - 80 - 90 - 100 - 110 - 120 - 130
140 - 150 - 160 - 170 - 180 - 190 - 200 - 210 - 220 - 230 - 240 - 250 - 260
270 - 280 - 290 - 300 - 310 - 320 - 330 - 340 - 350 - 360 - 370 - 380 - 390
400 - 410 - 420 - 430 - 440 - 450 - 460 - 470 - 480 - 490 - 500 - 510 - 520
530 - 540 - 550 - 560 - 570 - 580 - 590 - 600 - 610 - 620 - 630 - 640 - 650
660 - 670 - 680 - 690 - 700 - 710 - 720 - 730 - 740 - 750 - 760 - 770 - 780
790 - 800 - 810 - 820 - 830 - 840 - 850 - 860 - 870 - 880 - 890 - 900 - 910
920 - 930 - 940 - 950 - 960 - 970 - 980 - 990 - 1000
Notes: ____

____

____

____

Reg Reps: 10 - 20 - 30 - 40 - 50 - 60 - 70 - 80 - 90 - 100 - 110 - 120 - 130
140 - 150 - 160 - 170 - 180 - 190 - 200 - 210 - 220 - 230 - 240 - 250 - 260
270 - 280 - 290 - 300 - 310 - 320 - 330 - 340 - 350 - 360 - 370 - 380 - 390
400 - 410 - 420 - 430 - 440 - 450 - 460 - 470 - 480 - 490 - 500 - 510 - 520
530 - 540 - 550 - 560 - 570 - 580 - 590 - 600 - 610 - 620 - 630 - 640 - 650
660 - 670 - 680 - 690 - 700 - 710 - 720 - 730 - 740 - 750 - 760 - 770 - 780
790 - 800 - 810 - 820 - 830 - 840 - 850 - 860 - 870 - 880 - 890 - 900 - 910
920 - 930 - 940 - 950 - 960 - 970 - 980 - 990 - 1000

Notes: ______________________________

____

____

____

Reg Reps: 10 - 20 - 30 - 40 - 50 - 60 - 70 - 80 - 90 - 100 - 110 - 120 - 130
140 - 150 - 160 - 170 - 180 - 190 - 200 - 210 - 220 - 230 - 240 - 250 - 260
270 - 280 - 290 - 300 - 310 - 320 - 330 - 340 - 350 - 360 - 370 - 380 - 390
400 - 410 - 420 - 430 - 440 - 450 - 460 - 470 - 480 - 490 - 500 - 510 - 520
530 - 540 - 550 - 560 - 570 - 580 - 590 - 600 - 610 - 620 - 630 - 640 - 650
660 - 670 - 680 - 690 - 700 - 710 - 720 - 730 - 740 - 750 - 760 - 770 - 780
790 - 800 - 810 - 820 - 830 - 840 - 850 - 860 - 870 - 880 - 890 - 900 - 910
920 - 930 - 940 - 950 - 960 - 970 - 980 - 990 - 1000

Notes: ______________________________

____

____

____

Reg Reps: 10 - 20 - 30 - 40 - 50 - 60 - 70 - 80 - 90 - 100 - 110 - 120 - 130
140 - 150 - 160 - 170 - 180 - 190 - 200 - 210 - 220 - 230 - 240 - 250 - 260
270 - 280 - 290 - 300 - 310 - 320 - 330 - 340 - 350 - 360 - 370 - 380 - 390
400 - 410 - 420 - 430 - 440 - 450 - 460 - 470 - 480 - 490 - 500 - 510 - 520
530 - 540 - 550 - 560 - 570 - 580 - 590 - 600 - 610 - 620 - 630 - 640 - 650
660 - 670 - 680 - 690 - 700 - 710 - 720 - 730 - 740 - 750 - 760 - 770 - 780
790 - 800 - 810 - 820 - 830 - 840 - 850 - 860 - 870 - 880 - 890 - 900 - 910
920 - 930 - 940 - 950 - 960 - 970 - 980 - 990 - 1000

Notes: ______________________________

____

____

______________________________ **End Date:** ____________

Technique:__________________________ **L - R - Alt. Start Date:** _______

Notes:____

____

____

____

Slow Reps: 10 - 20 - 30 - 40 - 50 - 60 - 70 - 80 - 90 - 100 - 110 - 120 - 130
140 - 150 - 160 - 170 - 180 - 190 - 200 - 210 - 220 - 230 - 240 - 250 - 260
270 - 280 - 290 - 300 - 310 - 320 - 330 - 340 - 350 - 360 - 370 - 380 - 390
400 - 410 - 420 - 430 - 440 - 450 - 460 - 470 - 480 - 490 - 500 - 510 - 520
530 - 540 - 550 - 560 - 570 - 580 - 590 - 600 - 610 - 620 - 630 - 640 - 650
660 - 670 - 680 - 690 - 700 - 710 - 720 - 730 - 740 - 750 - 760 - 770 - 780
790 - 800 - 810 - 820 - 830 - 840 - 850 - 860 - 870 - 880 - 890 - 900 - 910
920 - 930 - 940 - 950 - 960 - 970 - 980 - 990 - 1000

Reg Reps: 10 - 20 - 30 - 40 - 50 - 60 - 70 - 80 - 90 - 100 - 110 - 120 - 130
140 - 150 - 160 - 170 - 180 - 190 - 200 - 210 - 220 - 230 - 240 - 250 - 260
270 - 280 - 290 - 300 - 310 - 320 - 330 - 340 - 350 - 360 - 370 - 380 - 390
400 - 410 - 420 - 430 - 440 - 450 - 460 - 470 - 480 - 490 - 500 - 510 - 520
530 - 540 - 550 - 560 - 570 - 580 - 590 - 600 - 610 - 620 - 630 - 640 - 650
660 - 670 - 680 - 690 - 700 - 710 - 720 - 730 - 740 - 750 - 760 - 770 - 780
790 - 800 - 810 - 820 - 830 - 840 - 850 - 860 - 870 - 880 - 890 - 900 - 910
920 - 930 - 940 - 950 - 960 - 970 - 980 - 990 - 1000

Notes: ____

____

____

____

Reg Reps: 10 - 20 - 30 - 40 - 50 - 60 - 70 - 80 - 90 - 100 - 110 - 120 - 130
140 - 150 - 160 - 170 - 180 - 190 - 200 - 210 - 220 - 230 - 240 - 250 - 260
270 - 280 - 290 - 300 - 310 - 320 - 330 - 340 - 350 - 360 - 370 - 380 - 390
400 - 410 - 420 - 430 - 440 - 450 - 460 - 470 - 480 - 490 - 500 - 510 - 520
530 - 540 - 550 - 560 - 570 - 580 - 590 - 600 - 610 - 620 - 630 - 640 - 650
660 - 670 - 680 - 690 - 700 - 710 - 720 - 730 - 740 - 750 - 760 - 770 - 780
790 - 800 - 810 - 820 - 830 - 840 - 850 - 860 - 870 - 880 - 890 - 900 - 910
920 - 930 - 940 - 950 - 960 - 970 - 980 - 990 - 1000

Notes: ____

____

____

____

Reg Reps: 10 - 20 - 30 - 40 - 50 - 60 - 70 - 80 - 90 - 100 - 110 - 120 - 130
140 - 150 - 160 - 170 - 180 - 190 - 200 - 210 - 220 - 230 - 240 - 250 - 260
270 - 280 - 290 - 300 - 310 - 320 - 330 - 340 - 350 - 360 - 370 - 380 - 390
400 - 410 - 420 - 430 - 440 - 450 - 460 - 470 - 480 - 490 - 500 - 510 - 520
530 - 540 - 550 - 560 - 570 - 580 - 590 - 600 - 610 - 620 - 630 - 640 - 650
660 - 670 - 680 - 690 - 700 - 710 - 720 - 730 - 740 - 750 - 760 - 770 - 780
790 - 800 - 810 - 820 - 830 - 840 - 850 - 860 - 870 - 880 - 890 - 900 - 910
920 - 930 - 940 - 950 - 960 - 970 - 980 - 990 - 1000

Notes: ____

____

____

____

Reg Reps: 10 - 20 - 30 - 40 - 50 - 60 - 70 - 80 - 90 - 100 - 110 - 120 - 130
140 - 150 - 160 - 170 - 180 - 190 - 200 - 210 - 220 - 230 - 240 - 250 - 260
270 - 280 - 290 - 300 - 310 - 320 - 330 - 340 - 350 - 360 - 370 - 380 - 390
400 - 410 - 420 - 430 - 440 - 450 - 460 - 470 - 480 - 490 - 500 - 510 - 520
530 - 540 - 550 - 560 - 570 - 580 - 590 - 600 - 610 - 620 - 630 - 640 - 650
660 - 670 - 680 - 690 - 700 - 710 - 720 - 730 - 740 - 750 - 760 - 770 - 780
790 - 800 - 810 - 820 - 830 - 840 - 850 - 860 - 870 - 880 - 890 - 900 - 910
920 - 930 - 940 - 950 - 960 - 970 - 980 - 990 - 1000

Notes: ____

____

____

____

Reg Reps: 10 - 20 - 30 - 40 - 50 - 60 - 70 - 80 - 90 - 100 - 110 - 120 - 130
140 - 150 - 160 - 170 - 180 - 190 - 200 - 210 - 220 - 230 - 240 - 250 - 260
270 - 280 - 290 - 300 - 310 - 320 - 330 - 340 - 350 - 360 - 370 - 380 - 390
400 - 410 - 420 - 430 - 440 - 450 - 460 - 470 - 480 - 490 - 500 - 510 - 520
530 - 540 - 550 - 560 - 570 - 580 - 590 - 600 - 610 - 620 - 630 - 640 - 650
660 - 670 - 680 - 690 - 700 - 710 - 720 - 730 - 740 - 750 - 760 - 770 - 780
790 - 800 - 810 - 820 - 830 - 840 - 850 - 860 - 870 - 880 - 890 - 900 - 910
920 - 930 - 940 - 950 - 960 - 970 - 980 - 990 - 1000

Notes: ____

____

____

________________________________ **End Date:** ____________

Technique: ______________________ **L - R - Alt. Start Date:** _______

Notes: ____

____

____

____

Slow Reps: 10 - 20 - 30 - 40 - 50 - 60 - 70 - 80 - 90 - 100 - 110 - 120 - 130
140 - 150 - 160 - 170 - 180 - 190 - 200 - 210 - 220 - 230 - 240 - 250 - 260
270 - 280 - 290 - 300 - 310 - 320 - 330 - 340 - 350 - 360 - 370 - 380 - 390
400 - 410 - 420 - 430 - 440 - 450 - 460 - 470 - 480 - 490 - 500 - 510 - 520
530 - 540 - 550 - 560 - 570 - 580 - 590 - 600 - 610 - 620 - 630 - 640 - 650
660 - 670 - 680 - 690 - 700 - 710 - 720 - 730 - 740 - 750 - 760 - 770 - 780
790 - 800 - 810 - 820 - 830 - 840 - 850 - 860 - 870 - 880 - 890 - 900 - 910
920 - 930 - 940 - 950 - 960 - 970 - 980 - 990 - 1000

Reg Reps: 10 - 20 - 30 - 40 - 50 - 60 - 70 - 80 - 90 - 100 - 110 - 120 - 130
140 - 150 - 160 - 170 - 180 - 190 - 200 - 210 - 220 - 230 - 240 - 250 - 260
270 - 280 - 290 - 300 - 310 - 320 - 330 - 340 - 350 - 360 - 370 - 380 - 390
400 - 410 - 420 - 430 - 440 - 450 - 460 - 470 - 480 - 490 - 500 - 510 - 520
530 - 540 - 550 - 560 - 570 - 580 - 590 - 600 - 610 - 620 - 630 - 640 - 650
660 - 670 - 680 - 690 - 700 - 710 - 720 - 730 - 740 - 750 - 760 - 770 - 780
790 - 800 - 810 - 820 - 830 - 840 - 850 - 860 - 870 - 880 - 890 - 900 - 910
920 - 930 - 940 - 950 - 960 - 970 - 980 - 990 - 1000

Notes: ____

____

____

____

Reg Reps: 10 - 20 - 30 - 40 - 50 - 60 - 70 - 80 - 90 - 100 - 110 - 120 - 130
140 - 150 - 160 - 170 - 180 - 190 - 200 - 210 - 220 - 230 - 240 - 250 - 260
270 - 280 - 290 - 300 - 310 - 320 - 330 - 340 - 350 - 360 - 370 - 380 - 390
400 - 410 - 420 - 430 - 440 - 450 - 460 - 470 - 480 - 490 - 500 - 510 - 520
530 - 540 - 550 - 560 - 570 - 580 - 590 - 600 - 610 - 620 - 630 - 640 - 650
660 - 670 - 680 - 690 - 700 - 710 - 720 - 730 - 740 - 750 - 760 - 770 - 780
790 - 800 - 810 - 820 - 830 - 840 - 850 - 860 - 870 - 880 - 890 - 900 - 910
920 - 930 - 940 - 950 - 960 - 970 - 980 - 990 - 1000

Notes: ____

____

____

____

Reg Reps: 10 - 20 - 30 - 40 - 50 - 60 - 70 - 80 - 90 - 100 - 110 - 120 - 130 - 140 - 150 - 160 - 170 - 180 - 190 - 200 - 210 - 220 - 230 - 240 - 250 - 260 - 270 - 280 - 290 - 300 - 310 - 320 - 330 - 340 - 350 - 360 - 370 - 380 - 390 - 400 - 410 - 420 - 430 - 440 - 450 - 460 - 470 - 480 - 490 - 500 - 510 - 520 - 530 - 540 - 550 - 560 - 570 - 580 - 590 - 600 - 610 - 620 - 630 - 640 - 650 - 660 - 670 - 680 - 690 - 700 - 710 - 720 - 730 - 740 - 750 - 760 - 770 - 780 - 790 - 800 - 810 - 820 - 830 - 840 - 850 - 860 - 870 - 880 - 890 - 900 - 910 - 920 - 930 - 940 - 950 - 960 - 970 - 980 - 990 - 1000

Notes: ______________________________

____

____

____

Reg Reps: 10 - 20 - 30 - 40 - 50 - 60 - 70 - 80 - 90 - 100 - 110 - 120 - 130 - 140 - 150 - 160 - 170 - 180 - 190 - 200 - 210 - 220 - 230 - 240 - 250 - 260 - 270 - 280 - 290 - 300 - 310 - 320 - 330 - 340 - 350 - 360 - 370 - 380 - 390 - 400 - 410 - 420 - 430 - 440 - 450 - 460 - 470 - 480 - 490 - 500 - 510 - 520 - 530 - 540 - 550 - 560 - 570 - 580 - 590 - 600 - 610 - 620 - 630 - 640 - 650 - 660 - 670 - 680 - 690 - 700 - 710 - 720 - 730 - 740 - 750 - 760 - 770 - 780 - 790 - 800 - 810 - 820 - 830 - 840 - 850 - 860 - 870 - 880 - 890 - 900 - 910 - 920 - 930 - 940 - 950 - 960 - 970 - 980 - 990 - 1000

Notes: ______________________________

____

____

____

Reg Reps: 10 - 20 - 30 - 40 - 50 - 60 - 70 - 80 - 90 - 100 - 110 - 120 - 130 - 140 - 150 - 160 - 170 - 180 - 190 - 200 - 210 - 220 - 230 - 240 - 250 - 260 - 270 - 280 - 290 - 300 - 310 - 320 - 330 - 340 - 350 - 360 - 370 - 380 - 390 - 400 - 410 - 420 - 430 - 440 - 450 - 460 - 470 - 480 - 490 - 500 - 510 - 520 - 530 - 540 - 550 - 560 - 570 - 580 - 590 - 600 - 610 - 620 - 630 - 640 - 650 - 660 - 670 - 680 - 690 - 700 - 710 - 720 - 730 - 740 - 750 - 760 - 770 - 780 - 790 - 800 - 810 - 820 - 830 - 840 - 850 - 860 - 870 - 880 - 890 - 900 - 910 - 920 - 930 - 940 - 950 - 960 - 970 - 980 - 990 - 1000

Notes: ______________________________

____

____

______________________________ **End Date:** ____________

Technique:________________ **L - R - Alt. Start Date:** _______

Notes:________________________

Slow Reps: 10 - 20 - 30 - 40 - 50 - 60 - 70 - 80 - 90 - 100 - 110 - 120 - 130
140 - 150 - 160 - 170 - 180 - 190 - 200 - 210 - 220 - 230 - 240 - 250 - 260
270 - 280 - 290 - 300 - 310 - 320 - 330 - 340 - 350 - 360 - 370 - 380 - 390
400 - 410 - 420 - 430 - 440 - 450 - 460 - 470 - 480 - 490 - 500 - 510 - 520
530 - 540 - 550 - 560 - 570 - 580 - 590 - 600 - 610 - 620 - 630 - 640 - 650
660 - 670 - 680 - 690 - 700 - 710 - 720 - 730 - 740 - 750 - 760 - 770 - 780
790 - 800 - 810 - 820 - 830 - 840 - 850 - 860 - 870 - 880 - 890 - 900 - 910
920 - 930 - 940 - 950 - 960 - 970 - 980 - 990 - 1000

Reg Reps: 10 - 20 - 30 - 40 - 50 - 60 - 70 - 80 - 90 - 100 - 110 - 120 - 130
140 - 150 - 160 - 170 - 180 - 190 - 200 - 210 - 220 - 230 - 240 - 250 - 260
270 - 280 - 290 - 300 - 310 - 320 - 330 - 340 - 350 - 360 - 370 - 380 - 390
400 - 410 - 420 - 430 - 440 - 450 - 460 - 470 - 480 - 490 - 500 - 510 - 520
530 - 540 - 550 - 560 - 570 - 580 - 590 - 600 - 610 - 620 - 630 - 640 - 650
660 - 670 - 680 - 690 - 700 - 710 - 720 - 730 - 740 - 750 - 760 - 770 - 780
790 - 800 - 810 - 820 - 830 - 840 - 850 - 860 - 870 - 880 - 890 - 900 - 910
920 - 930 - 940 - 950 - 960 - 970 - 980 - 990 - 1000

Notes: ________________________

Reg Reps: 10 - 20 - 30 - 40 - 50 - 60 - 70 - 80 - 90 - 100 - 110 - 120 - 130
140 - 150 - 160 - 170 - 180 - 190 - 200 - 210 - 220 - 230 - 240 - 250 - 260
270 - 280 - 290 - 300 - 310 - 320 - 330 - 340 - 350 - 360 - 370 - 380 - 390
400 - 410 - 420 - 430 - 440 - 450 - 460 - 470 - 480 - 490 - 500 - 510 - 520
530 - 540 - 550 - 560 - 570 - 580 - 590 - 600 - 610 - 620 - 630 - 640 - 650
660 - 670 - 680 - 690 - 700 - 710 - 720 - 730 - 740 - 750 - 760 - 770 - 780
790 - 800 - 810 - 820 - 830 - 840 - 850 - 860 - 870 - 880 - 890 - 900 - 910
920 - 930 - 940 - 950 - 960 - 970 - 980 - 990 - 1000

Notes: ________________________

Reg Reps: 10 - 20 - 30 - 40 - 50 - 60 - 70 - 80 - 90 - 100 - 110 - 120 - 130
140 - 150 - 160 - 170 - 180 - 190 - 200 - 210 - 220 - 230 - 240 - 250 - 260
270 - 280 - 290 - 300 - 310 - 320 - 330 - 340 - 350 - 360 - 370 - 380 - 390
400 - 410 - 420 - 430 - 440 - 450 - 460 - 470 - 480 - 490 - 500 - 510 - 520
530 - 540 - 550 - 560 - 570 - 580 - 590 - 600 - 610 - 620 - 630 - 640 - 650
660 - 670 - 680 - 690 - 700 - 710 - 720 - 730 - 740 - 750 - 760 - 770 - 780
790 - 800 - 810 - 820 - 830 - 840 - 850 - 860 - 870 - 880 - 890 - 900 - 910
920 - 930 - 940 - 950 - 960 - 970 - 980 - 990 - 1000

Notes: ____

____

____

____

Reg Reps: 10 - 20 - 30 - 40 - 50 - 60 - 70 - 80 - 90 - 100 - 110 - 120 - 130
140 - 150 - 160 - 170 - 180 - 190 - 200 - 210 - 220 - 230 - 240 - 250 - 260
270 - 280 - 290 - 300 - 310 - 320 - 330 - 340 - 350 - 360 - 370 - 380 - 390
400 - 410 - 420 - 430 - 440 - 450 - 460 - 470 - 480 - 490 - 500 - 510 - 520
530 - 540 - 550 - 560 - 570 - 580 - 590 - 600 - 610 - 620 - 630 - 640 - 650
660 - 670 - 680 - 690 - 700 - 710 - 720 - 730 - 740 - 750 - 760 - 770 - 780
790 - 800 - 810 - 820 - 830 - 840 - 850 - 860 - 870 - 880 - 890 - 900 - 910
920 - 930 - 940 - 950 - 960 - 970 - 980 - 990 - 1000

Notes: ____

____

____

____

Reg Reps: 10 - 20 - 30 - 40 - 50 - 60 - 70 - 80 - 90 - 100 - 110 - 120 - 130
140 - 150 - 160 - 170 - 180 - 190 - 200 - 210 - 220 - 230 - 240 - 250 - 260
270 - 280 - 290 - 300 - 310 - 320 - 330 - 340 - 350 - 360 - 370 - 380 - 390
400 - 410 - 420 - 430 - 440 - 450 - 460 - 470 - 480 - 490 - 500 - 510 - 520
530 - 540 - 550 - 560 - 570 - 580 - 590 - 600 - 610 - 620 - 630 - 640 - 650
660 - 670 - 680 - 690 - 700 - 710 - 720 - 730 - 740 - 750 - 760 - 770 - 780
790 - 800 - 810 - 820 - 830 - 840 - 850 - 860 - 870 - 880 - 890 - 900 - 910
920 - 930 - 940 - 950 - 960 - 970 - 980 - 990 - 1000

Notes: ____

____

____

______________________________ **End Date:** __________

Technique:__________________________ **L - R - Alt. Start Date:** ________
Notes:____

Slow Reps: 10 - 20 - 30 - 40 - 50 - 60 - 70 - 80 - 90 - 100 - 110 - 120 - 130
140 - 150 - 160 - 170 - 180 - 190 - 200 - 210 - 220 - 230 - 240 - 250 - 260
270 - 280 - 290 - 300 - 310 - 320 - 330 - 340 - 350 - 360 - 370 - 380 - 390
400 - 410 - 420 - 430 - 440 - 450 - 460 - 470 - 480 - 490 - 500 - 510 - 520
530 - 540 - 550 - 560 - 570 - 580 - 590 - 600 - 610 - 620 - 630 - 640 - 650
660 - 670 - 680 - 690 - 700 - 710 - 720 - 730 - 740 - 750 - 760 - 770 - 780
790 - 800 - 810 - 820 - 830 - 840 - 850 - 860 - 870 - 880 - 890 - 900 - 910
920 - 930 - 940 - 950 - 960 - 970 - 980 - 990 - 1000

Reg Reps: 10 - 20 - 30 - 40 - 50 - 60 - 70 - 80 - 90 - 100 - 110 - 120 - 130
140 - 150 - 160 - 170 - 180 - 190 - 200 - 210 - 220 - 230 - 240 - 250 - 260
270 - 280 - 290 - 300 - 310 - 320 - 330 - 340 - 350 - 360 - 370 - 380 - 390
400 - 410 - 420 - 430 - 440 - 450 - 460 - 470 - 480 - 490 - 500 - 510 - 520
530 - 540 - 550 - 560 - 570 - 580 - 590 - 600 - 610 - 620 - 630 - 640 - 650
660 - 670 - 680 - 690 - 700 - 710 - 720 - 730 - 740 - 750 - 760 - 770 - 780
790 - 800 - 810 - 820 - 830 - 840 - 850 - 860 - 870 - 880 - 890 - 900 - 910
920 - 930 - 940 - 950 - 960 - 970 - 980 - 990 - 1000
Notes: ____

Reg Reps: 10 - 20 - 30 - 40 - 50 - 60 - 70 - 80 - 90 - 100 - 110 - 120 - 130
140 - 150 - 160 - 170 - 180 - 190 - 200 - 210 - 220 - 230 - 240 - 250 - 260
270 - 280 - 290 - 300 - 310 - 320 - 330 - 340 - 350 - 360 - 370 - 380 - 390
400 - 410 - 420 - 430 - 440 - 450 - 460 - 470 - 480 - 490 - 500 - 510 - 520
530 - 540 - 550 - 560 - 570 - 580 - 590 - 600 - 610 - 620 - 630 - 640 - 650
660 - 670 - 680 - 690 - 700 - 710 - 720 - 730 - 740 - 750 - 760 - 770 - 780
790 - 800 - 810 - 820 - 830 - 840 - 850 - 860 - 870 - 880 - 890 - 900 - 910
920 - 930 - 940 - 950 - 960 - 970 - 980 - 990 - 1000
Notes: ____

Reg Reps: 10 - 20 - 30 - 40 - 50 - 60 - 70 - 80 - 90 - 100 - 110 - 120 - 130
140 - 150 - 160 - 170 - 180 - 190 - 200 - 210 - 220 - 230 - 240 - 250 - 260
270 - 280 - 290 - 300 - 310 - 320 - 330 - 340 - 350 - 360 - 370 - 380 - 390
400 - 410 - 420 - 430 - 440 - 450 - 460 - 470 - 480 - 490 - 500 - 510 - 520
530 - 540 - 550 - 560 - 570 - 580 - 590 - 600 - 610 - 620 - 630 - 640 - 650
660 - 670 - 680 - 690 - 700 - 710 - 720 - 730 - 740 - 750 - 760 - 770 - 780
790 - 800 - 810 - 820 - 830 - 840 - 850 - 860 - 870 - 880 - 890 - 900 - 910
920 - 930 - 940 - 950 - 960 - 970 - 980 - 990 - 1000

Notes: ______________________________

____

____

____

Reg Reps: 10 - 20 - 30 - 40 - 50 - 60 - 70 - 80 - 90 - 100 - 110 - 120 - 130
140 - 150 - 160 - 170 - 180 - 190 - 200 - 210 - 220 - 230 - 240 - 250 - 260
270 - 280 - 290 - 300 - 310 - 320 - 330 - 340 - 350 - 360 - 370 - 380 - 390
400 - 410 - 420 - 430 - 440 - 450 - 460 - 470 - 480 - 490 - 500 - 510 - 520
530 - 540 - 550 - 560 - 570 - 580 - 590 - 600 - 610 - 620 - 630 - 640 - 650
660 - 670 - 680 - 690 - 700 - 710 - 720 - 730 - 740 - 750 - 760 - 770 - 780
790 - 800 - 810 - 820 - 830 - 840 - 850 - 860 - 870 - 880 - 890 - 900 - 910
920 - 930 - 940 - 950 - 960 - 970 - 980 - 990 - 1000

Notes: ______________________________

____

____

____

Reg Reps: 10 - 20 - 30 - 40 - 50 - 60 - 70 - 80 - 90 - 100 - 110 - 120 - 130
140 - 150 - 160 - 170 - 180 - 190 - 200 - 210 - 220 - 230 - 240 - 250 - 260
270 - 280 - 290 - 300 - 310 - 320 - 330 - 340 - 350 - 360 - 370 - 380 - 390
400 - 410 - 420 - 430 - 440 - 450 - 460 - 470 - 480 - 490 - 500 - 510 - 520
530 - 540 - 550 - 560 - 570 - 580 - 590 - 600 - 610 - 620 - 630 - 640 - 650
660 - 670 - 680 - 690 - 700 - 710 - 720 - 730 - 740 - 750 - 760 - 770 - 780
790 - 800 - 810 - 820 - 830 - 840 - 850 - 860 - 870 - 880 - 890 - 900 - 910
920 - 930 - 940 - 950 - 960 - 970 - 980 - 990 - 1000

Notes: ______________________________

____

____

______________________________ **End Date:** ____________

Technique:________________________ **L - R - Alt. Start Date:** _______

Notes:____

Slow Reps: 10 - 20 - 30 - 40 - 50 - 60 - 70 - 80 - 90 - 100 - 110 - 120 - 130
140 - 150 - 160 - 170 - 180 - 190 - 200 - 210 - 220 - 230 - 240 - 250 - 260
270 - 280 - 290 - 300 - 310 - 320 - 330 - 340 - 350 - 360 - 370 - 380 - 390
400 - 410 - 420 - 430 - 440 - 450 - 460 - 470 - 480 - 490 - 500 - 510 - 520
530 - 540 - 550 - 560 - 570 - 580 - 590 - 600 - 610 - 620 - 630 - 640 - 650
660 - 670 - 680 - 690 - 700 - 710 - 720 - 730 - 740 - 750 - 760 - 770 - 780
790 - 800 - 810 - 820 - 830 - 840 - 850 - 860 - 870 - 880 - 890 - 900 - 910
920 - 930 - 940 - 950 - 960 - 970 - 980 - 990 - 1000

Reg Reps: 10 - 20 - 30 - 40 - 50 - 60 - 70 - 80 - 90 - 100 - 110 - 120 - 130
140 - 150 - 160 - 170 - 180 - 190 - 200 - 210 - 220 - 230 - 240 - 250 - 260
270 - 280 - 290 - 300 - 310 - 320 - 330 - 340 - 350 - 360 - 370 - 380 - 390
400 - 410 - 420 - 430 - 440 - 450 - 460 - 470 - 480 - 490 - 500 - 510 - 520
530 - 540 - 550 - 560 - 570 - 580 - 590 - 600 - 610 - 620 - 630 - 640 - 650
660 - 670 - 680 - 690 - 700 - 710 - 720 - 730 - 740 - 750 - 760 - 770 - 780
790 - 800 - 810 - 820 - 830 - 840 - 850 - 860 - 870 - 880 - 890 - 900 - 910
920 - 930 - 940 - 950 - 960 - 970 - 980 - 990 - 1000

Notes: ____

Reg Reps: 10 - 20 - 30 - 40 - 50 - 60 - 70 - 80 - 90 - 100 - 110 - 120 - 130
140 - 150 - 160 - 170 - 180 - 190 - 200 - 210 - 220 - 230 - 240 - 250 - 260
270 - 280 - 290 - 300 - 310 - 320 - 330 - 340 - 350 - 360 - 370 - 380 - 390
400 - 410 - 420 - 430 - 440 - 450 - 460 - 470 - 480 - 490 - 500 - 510 - 520
530 - 540 - 550 - 560 - 570 - 580 - 590 - 600 - 610 - 620 - 630 - 640 - 650
660 - 670 - 680 - 690 - 700 - 710 - 720 - 730 - 740 - 750 - 760 - 770 - 780
790 - 800 - 810 - 820 - 830 - 840 - 850 - 860 - 870 - 880 - 890 - 900 - 910
920 - 930 - 940 - 950 - 960 - 970 - 980 - 990 - 1000

Notes: ____

Reg Reps: 10 - 20 - 30 - 40 - 50 - 60 - 70 - 80 - 90 - 100 - 110 - 120 - 130
140 - 150 - 160 - 170 - 180 - 190 - 200 - 210 - 220 - 230 - 240 - 250 - 260
270 - 280 - 290 - 300 - 310 - 320 - 330 - 340 - 350 - 360 - 370 - 380 - 390
400 - 410 - 420 - 430 - 440 - 450 - 460 - 470 - 480 - 490 - 500 - 510 - 520
530 - 540 - 550 - 560 - 570 - 580 - 590 - 600 - 610 - 620 - 630 - 640 - 650
660 - 670 - 680 - 690 - 700 - 710 - 720 - 730 - 740 - 750 - 760 - 770 - 780
790 - 800 - 810 - 820 - 830 - 840 - 850 - 860 - 870 - 880 - 890 - 900 - 910
920 - 930 - 940 - 950 - 960 - 970 - 980 - 990 - 1000

Notes: ____

____

____

____

Reg Reps: 10 - 20 - 30 - 40 - 50 - 60 - 70 - 80 - 90 - 100 - 110 - 120 - 130
140 - 150 - 160 - 170 - 180 - 190 - 200 - 210 - 220 - 230 - 240 - 250 - 260
270 - 280 - 290 - 300 - 310 - 320 - 330 - 340 - 350 - 360 - 370 - 380 - 390
400 - 410 - 420 - 430 - 440 - 450 - 460 - 470 - 480 - 490 - 500 - 510 - 520
530 - 540 - 550 - 560 - 570 - 580 - 590 - 600 - 610 - 620 - 630 - 640 - 650
660 - 670 - 680 - 690 - 700 - 710 - 720 - 730 - 740 - 750 - 760 - 770 - 780
790 - 800 - 810 - 820 - 830 - 840 - 850 - 860 - 870 - 880 - 890 - 900 - 910
920 - 930 - 940 - 950 - 960 - 970 - 980 - 990 - 1000

Notes: ____

____

____

____

Reg Reps: 10 - 20 - 30 - 40 - 50 - 60 - 70 - 80 - 90 - 100 - 110 - 120 - 130
140 - 150 - 160 - 170 - 180 - 190 - 200 - 210 - 220 - 230 - 240 - 250 - 260
270 - 280 - 290 - 300 - 310 - 320 - 330 - 340 - 350 - 360 - 370 - 380 - 390
400 - 410 - 420 - 430 - 440 - 450 - 460 - 470 - 480 - 490 - 500 - 510 - 520
530 - 540 - 550 - 560 - 570 - 580 - 590 - 600 - 610 - 620 - 630 - 640 - 650
660 - 670 - 680 - 690 - 700 - 710 - 720 - 730 - 740 - 750 - 760 - 770 - 780
790 - 800 - 810 - 820 - 830 - 840 - 850 - 860 - 870 - 880 - 890 - 900 - 910
920 - 930 - 940 - 950 - 960 - 970 - 980 - 990 - 1000

Notes: ____

____

____

________________________________ **End Date:** ____________

Technique:____________________ **L - R - Alt. Start Date:** _______
Notes:____

Slow Reps: 10 - 20 - 30 - 40 - 50 - 60 - 70 - 80 - 90 - 100 - 110 - 120 - 130
140 - 150 - 160 - 170 - 180 - 190 - 200 - 210 - 220 - 230 - 240 - 250 - 260
270 - 280 - 290 - 300 - 310 - 320 - 330 - 340 - 350 - 360 - 370 - 380 - 390
400 - 410 - 420 - 430 - 440 - 450 - 460 - 470 - 480 - 490 - 500 - 510 - 520
530 - 540 - 550 - 560 - 570 - 580 - 590 - 600 - 610 - 620 - 630 - 640 - 650
660 - 670 - 680 - 690 - 700 - 710 - 720 - 730 - 740 - 750 - 760 - 770 - 780
790 - 800 - 810 - 820 - 830 - 840 - 850 - 860 - 870 - 880 - 890 - 900 - 910
920 - 930 - 940 - 950 - 960 - 970 - 980 - 990 - 1000

Reg Reps: 10 - 20 - 30 - 40 - 50 - 60 - 70 - 80 - 90 - 100 - 110 - 120 - 130
140 - 150 - 160 - 170 - 180 - 190 - 200 - 210 - 220 - 230 - 240 - 250 - 260
270 - 280 - 290 - 300 - 310 - 320 - 330 - 340 - 350 - 360 - 370 - 380 - 390
400 - 410 - 420 - 430 - 440 - 450 - 460 - 470 - 480 - 490 - 500 - 510 - 520
530 - 540 - 550 - 560 - 570 - 580 - 590 - 600 - 610 - 620 - 630 - 640 - 650
660 - 670 - 680 - 690 - 700 - 710 - 720 - 730 - 740 - 750 - 760 - 770 - 780
790 - 800 - 810 - 820 - 830 - 840 - 850 - 860 - 870 - 880 - 890 - 900 - 910
920 - 930 - 940 - 950 - 960 - 970 - 980 - 990 - 1000
Notes: ____

Reg Reps: 10 - 20 - 30 - 40 - 50 - 60 - 70 - 80 - 90 - 100 - 110 - 120 - 130
140 - 150 - 160 - 170 - 180 - 190 - 200 - 210 - 220 - 230 - 240 - 250 - 260
270 - 280 - 290 - 300 - 310 - 320 - 330 - 340 - 350 - 360 - 370 - 380 - 390
400 - 410 - 420 - 430 - 440 - 450 - 460 - 470 - 480 - 490 - 500 - 510 - 520
530 - 540 - 550 - 560 - 570 - 580 - 590 - 600 - 610 - 620 - 630 - 640 - 650
660 - 670 - 680 - 690 - 700 - 710 - 720 - 730 - 740 - 750 - 760 - 770 - 780
790 - 800 - 810 - 820 - 830 - 840 - 850 - 860 - 870 - 880 - 890 - 900 - 910
920 - 930 - 940 - 950 - 960 - 970 - 980 - 990 - 1000
Notes: ____

Reg Reps: 10 - 20 - 30 - 40 - 50 - 60 - 70 - 80 - 90 - 100 - 110 - 120 - 130
140 - 150 - 160 - 170 - 180 - 190 - 200 - 210 - 220 - 230 - 240 - 250 - 260
270 - 280 - 290 - 300 - 310 - 320 - 330 - 340 - 350 - 360 - 370 - 380 - 390
400 - 410 - 420 - 430 - 440 - 450 - 460 - 470 - 480 - 490 - 500 - 510 - 520
530 - 540 - 550 - 560 - 570 - 580 - 590 - 600 - 610 - 620 - 630 - 640 - 650
660 - 670 - 680 - 690 - 700 - 710 - 720 - 730 - 740 - 750 - 760 - 770 - 780
790 - 800 - 810 - 820 - 830 - 840 - 850 - 860 - 870 - 880 - 890 - 900 - 910
920 - 930 - 940 - 950 - 960 - 970 - 980 - 990 - 1000

Notes: ______________________________

____

____

____

Reg Reps: 10 - 20 - 30 - 40 - 50 - 60 - 70 - 80 - 90 - 100 - 110 - 120 - 130
140 - 150 - 160 - 170 - 180 - 190 - 200 - 210 - 220 - 230 - 240 - 250 - 260
270 - 280 - 290 - 300 - 310 - 320 - 330 - 340 - 350 - 360 - 370 - 380 - 390
400 - 410 - 420 - 430 - 440 - 450 - 460 - 470 - 480 - 490 - 500 - 510 - 520
530 - 540 - 550 - 560 - 570 - 580 - 590 - 600 - 610 - 620 - 630 - 640 - 650
660 - 670 - 680 - 690 - 700 - 710 - 720 - 730 - 740 - 750 - 760 - 770 - 780
790 - 800 - 810 - 820 - 830 - 840 - 850 - 860 - 870 - 880 - 890 - 900 - 910
920 - 930 - 940 - 950 - 960 - 970 - 980 - 990 - 1000

Notes: ______________________________

____

____

____

Reg Reps: 10 - 20 - 30 - 40 - 50 - 60 - 70 - 80 - 90 - 100 - 110 - 120 - 130
140 - 150 - 160 - 170 - 180 - 190 - 200 - 210 - 220 - 230 - 240 - 250 - 260
270 - 280 - 290 - 300 - 310 - 320 - 330 - 340 - 350 - 360 - 370 - 380 - 390
400 - 410 - 420 - 430 - 440 - 450 - 460 - 470 - 480 - 490 - 500 - 510 - 520
530 - 540 - 550 - 560 - 570 - 580 - 590 - 600 - 610 - 620 - 630 - 640 - 650
660 - 670 - 680 - 690 - 700 - 710 - 720 - 730 - 740 - 750 - 760 - 770 - 780
790 - 800 - 810 - 820 - 830 - 840 - 850 - 860 - 870 - 880 - 890 - 900 - 910
920 - 930 - 940 - 950 - 960 - 970 - 980 - 990 - 1000

Notes: ______________________________

____

____

______________________________ **End Date:** ______________

Technique:________________________ **L - R - Alt. Start Date:** ________
Notes:____

Slow Reps: 10 - 20 - 30 - 40 - 50 - 60 - 70 - 80 - 90 - 100 - 110 - 120 - 130
140 - 150 - 160 - 170 - 180 - 190 - 200 - 210 - 220 - 230 - 240 - 250 - 260
270 - 280 - 290 - 300 - 310 - 320 - 330 - 340 - 350 - 360 - 370 - 380 - 390
400 - 410 - 420 - 430 - 440 - 450 - 460 - 470 - 480 - 490 - 500 - 510 - 520
530 - 540 - 550 - 560 - 570 - 580 - 590 - 600 - 610 - 620 - 630 - 640 - 650
660 - 670 - 680 - 690 - 700 - 710 - 720 - 730 - 740 - 750 - 760 - 770 - 780
790 - 800 - 810 - 820 - 830 - 840 - 850 - 860 - 870 - 880 - 890 - 900 - 910
920 - 930 - 940 - 950 - 960 - 970 - 980 - 990 - 1000

Reg Reps: 10 - 20 - 30 - 40 - 50 - 60 - 70 - 80 - 90 - 100 - 110 - 120 - 130
140 - 150 - 160 - 170 - 180 - 190 - 200 - 210 - 220 - 230 - 240 - 250 - 260
270 - 280 - 290 - 300 - 310 - 320 - 330 - 340 - 350 - 360 - 370 - 380 - 390
400 - 410 - 420 - 430 - 440 - 450 - 460 - 470 - 480 - 490 - 500 - 510 - 520
530 - 540 - 550 - 560 - 570 - 580 - 590 - 600 - 610 - 620 - 630 - 640 - 650
660 - 670 - 680 - 690 - 700 - 710 - 720 - 730 - 740 - 750 - 760 - 770 - 780
790 - 800 - 810 - 820 - 830 - 840 - 850 - 860 - 870 - 880 - 890 - 900 - 910
920 - 930 - 940 - 950 - 960 - 970 - 980 - 990 - 1000
Notes: ____

Reg Reps: 10 - 20 - 30 - 40 - 50 - 60 - 70 - 80 - 90 - 100 - 110 - 120 - 130
140 - 150 - 160 - 170 - 180 - 190 - 200 - 210 - 220 - 230 - 240 - 250 - 260
270 - 280 - 290 - 300 - 310 - 320 - 330 - 340 - 350 - 360 - 370 - 380 - 390
400 - 410 - 420 - 430 - 440 - 450 - 460 - 470 - 480 - 490 - 500 - 510 - 520
530 - 540 - 550 - 560 - 570 - 580 - 590 - 600 - 610 - 620 - 630 - 640 - 650
660 - 670 - 680 - 690 - 700 - 710 - 720 - 730 - 740 - 750 - 760 - 770 - 780
790 - 800 - 810 - 820 - 830 - 840 - 850 - 860 - 870 - 880 - 890 - 900 - 910
920 - 930 - 940 - 950 - 960 - 970 - 980 - 990 - 1000
Notes: ____

Reg Reps: 10 - 20 - 30 - 40 - 50 - 60 - 70 - 80 - 90 - 100 - 110 - 120 - 130
140 - 150 - 160 - 170 - 180 - 190 - 200 - 210 - 220 - 230 - 240 - 250 - 260
270 - 280 - 290 - 300 - 310 - 320 - 330 - 340 - 350 - 360 - 370 - 380 - 390
400 - 410 - 420 - 430 - 440 - 450 - 460 - 470 - 480 - 490 - 500 - 510 - 520
530 - 540 - 550 - 560 - 570 - 580 - 590 - 600 - 610 - 620 - 630 - 640 - 650
660 - 670 - 680 - 690 - 700 - 710 - 720 - 730 - 740 - 750 - 760 - 770 - 780
790 - 800 - 810 - 820 - 830 - 840 - 850 - 860 - 870 - 880 - 890 - 900 - 910
920 - 930 - 940 - 950 - 960 - 970 - 980 - 990 - 1000

Notes: ____

____

____

____

Reg Reps: 10 - 20 - 30 - 40 - 50 - 60 - 70 - 80 - 90 - 100 - 110 - 120 - 130
140 - 150 - 160 - 170 - 180 - 190 - 200 - 210 - 220 - 230 - 240 - 250 - 260
270 - 280 - 290 - 300 - 310 - 320 - 330 - 340 - 350 - 360 - 370 - 380 - 390
400 - 410 - 420 - 430 - 440 - 450 - 460 - 470 - 480 - 490 - 500 - 510 - 520
530 - 540 - 550 - 560 - 570 - 580 - 590 - 600 - 610 - 620 - 630 - 640 - 650
660 - 670 - 680 - 690 - 700 - 710 - 720 - 730 - 740 - 750 - 760 - 770 - 780
790 - 800 - 810 - 820 - 830 - 840 - 850 - 860 - 870 - 880 - 890 - 900 - 910
920 - 930 - 940 - 950 - 960 - 970 - 980 - 990 - 1000

Notes: ____

____

____

____

Reg Reps: 10 - 20 - 30 - 40 - 50 - 60 - 70 - 80 - 90 - 100 - 110 - 120 - 130
140 - 150 - 160 - 170 - 180 - 190 - 200 - 210 - 220 - 230 - 240 - 250 - 260
270 - 280 - 290 - 300 - 310 - 320 - 330 - 340 - 350 - 360 - 370 - 380 - 390
400 - 410 - 420 - 430 - 440 - 450 - 460 - 470 - 480 - 490 - 500 - 510 - 520
530 - 540 - 550 - 560 - 570 - 580 - 590 - 600 - 610 - 620 - 630 - 640 - 650
660 - 670 - 680 - 690 - 700 - 710 - 720 - 730 - 740 - 750 - 760 - 770 - 780
790 - 800 - 810 - 820 - 830 - 840 - 850 - 860 - 870 - 880 - 890 - 900 - 910
920 - 930 - 940 - 950 - 960 - 970 - 980 - 990 - 1000

Notes: ____

____

____

______________________________ **End Date:** ____________

Technique:________________________ **L - R - Alt. Start Date:** ________

Notes:____

Slow Reps: 10 - 20 - 30 - 40 - 50 - 60 - 70 - 80 - 90 - 100 - 110 - 120 - 130
140 - 150 - 160 - 170 - 180 - 190 - 200 - 210 - 220 - 230 - 240 - 250 - 260
270 - 280 - 290 - 300 - 310 - 320 - 330 - 340 - 350 - 360 - 370 - 380 - 390
400 - 410 - 420 - 430 - 440 - 450 - 460 - 470 - 480 - 490 - 500 - 510 - 520
530 - 540 - 550 - 560 - 570 - 580 - 590 - 600 - 610 - 620 - 630 - 640 - 650
660 - 670 - 680 - 690 - 700 - 710 - 720 - 730 - 740 - 750 - 760 - 770 - 780
790 - 800 - 810 - 820 - 830 - 840 - 850 - 860 - 870 - 880 - 890 - 900 - 910
920 - 930 - 940 - 950 - 960 - 970 - 980 - 990 - 1000

Reg Reps: 10 - 20 - 30 - 40 - 50 - 60 - 70 - 80 - 90 - 100 - 110 - 120 - 130
140 - 150 - 160 - 170 - 180 - 190 - 200 - 210 - 220 - 230 - 240 - 250 - 260
270 - 280 - 290 - 300 - 310 - 320 - 330 - 340 - 350 - 360 - 370 - 380 - 390
400 - 410 - 420 - 430 - 440 - 450 - 460 - 470 - 480 - 490 - 500 - 510 - 520
530 - 540 - 550 - 560 - 570 - 580 - 590 - 600 - 610 - 620 - 630 - 640 - 650
660 - 670 - 680 - 690 - 700 - 710 - 720 - 730 - 740 - 750 - 760 - 770 - 780
790 - 800 - 810 - 820 - 830 - 840 - 850 - 860 - 870 - 880 - 890 - 900 - 910
920 - 930 - 940 - 950 - 960 - 970 - 980 - 990 - 1000

Notes: ____

Reg Reps: 10 - 20 - 30 - 40 - 50 - 60 - 70 - 80 - 90 - 100 - 110 - 120 - 130
140 - 150 - 160 - 170 - 180 - 190 - 200 - 210 - 220 - 230 - 240 - 250 - 260
270 - 280 - 290 - 300 - 310 - 320 - 330 - 340 - 350 - 360 - 370 - 380 - 390
400 - 410 - 420 - 430 - 440 - 450 - 460 - 470 - 480 - 490 - 500 - 510 - 520
530 - 540 - 550 - 560 - 570 - 580 - 590 - 600 - 610 - 620 - 630 - 640 - 650
660 - 670 - 680 - 690 - 700 - 710 - 720 - 730 - 740 - 750 - 760 - 770 - 780
790 - 800 - 810 - 820 - 830 - 840 - 850 - 860 - 870 - 880 - 890 - 900 - 910
920 - 930 - 940 - 950 - 960 - 970 - 980 - 990 - 1000

Notes: ____

Reg Reps: 10 - 20 - 30 - 40 - 50 - 60 - 70 - 80 - 90 - 100 - 110 - 120 - 130
140 - 150 - 160 - 170 - 180 - 190 - 200 - 210 - 220 - 230 - 240 - 250 - 260
270 - 280 - 290 - 300 - 310 - 320 - 330 - 340 - 350 - 360 - 370 - 380 - 390
400 - 410 - 420 - 430 - 440 - 450 - 460 - 470 - 480 - 490 - 500 - 510 - 520
530 - 540 - 550 - 560 - 570 - 580 - 590 - 600 - 610 - 620 - 630 - 640 - 650
660 - 670 - 680 - 690 - 700 - 710 - 720 - 730 - 740 - 750 - 760 - 770 - 780
790 - 800 - 810 - 820 - 830 - 840 - 850 - 860 - 870 - 880 - 890 - 900 - 910
920 - 930 - 940 - 950 - 960 - 970 - 980 - 990 - 1000

Notes: ____

____

____

____

Reg Reps: 10 - 20 - 30 - 40 - 50 - 60 - 70 - 80 - 90 - 100 - 110 - 120 - 130
140 - 150 - 160 - 170 - 180 - 190 - 200 - 210 - 220 - 230 - 240 - 250 - 260
270 - 280 - 290 - 300 - 310 - 320 - 330 - 340 - 350 - 360 - 370 - 380 - 390
400 - 410 - 420 - 430 - 440 - 450 - 460 - 470 - 480 - 490 - 500 - 510 - 520
530 - 540 - 550 - 560 - 570 - 580 - 590 - 600 - 610 - 620 - 630 - 640 - 650
660 - 670 - 680 - 690 - 700 - 710 - 720 - 730 - 740 - 750 - 760 - 770 - 780
790 - 800 - 810 - 820 - 830 - 840 - 850 - 860 - 870 - 880 - 890 - 900 - 910
920 - 930 - 940 - 950 - 960 - 970 - 980 - 990 - 1000

Notes: ____

____

____

____

Reg Reps: 10 - 20 - 30 - 40 - 50 - 60 - 70 - 80 - 90 - 100 - 110 - 120 - 130
140 - 150 - 160 - 170 - 180 - 190 - 200 - 210 - 220 - 230 - 240 - 250 - 260
270 - 280 - 290 - 300 - 310 - 320 - 330 - 340 - 350 - 360 - 370 - 380 - 390
400 - 410 - 420 - 430 - 440 - 450 - 460 - 470 - 480 - 490 - 500 - 510 - 520
530 - 540 - 550 - 560 - 570 - 580 - 590 - 600 - 610 - 620 - 630 - 640 - 650
660 - 670 - 680 - 690 - 700 - 710 - 720 - 730 - 740 - 750 - 760 - 770 - 780
790 - 800 - 810 - 820 - 830 - 840 - 850 - 860 - 870 - 880 - 890 - 900 - 910
920 - 930 - 940 - 950 - 960 - 970 - 980 - 990 - 1000

Notes: ____

____

____

______________________________ **End Date:** ______________

Technique:_________________________ **L - R - Alt. Start Date:** _______

Notes:____

____

____

____

Slow Reps: 10 - 20 - 30 - 40 - 50 - 60 - 70 - 80 - 90 - 100 - 110 - 120 - 130
140 - 150 - 160 - 170 - 180 - 190 - 200 - 210 - 220 - 230 - 240 - 250 - 260
270 - 280 - 290 - 300 - 310 - 320 - 330 - 340 - 350 - 360 - 370 - 380 - 390
400 - 410 - 420 - 430 - 440 - 450 - 460 - 470 - 480 - 490 - 500 - 510 - 520
530 - 540 - 550 - 560 - 570 - 580 - 590 - 600 - 610 - 620 - 630 - 640 - 650
660 - 670 - 680 - 690 - 700 - 710 - 720 - 730 - 740 - 750 - 760 - 770 - 780
790 - 800 - 810 - 820 - 830 - 840 - 850 - 860 - 870 - 880 - 890 - 900 - 910
920 - 930 - 940 - 950 - 960 - 970 - 980 - 990 - 1000

Reg Reps: 10 - 20 - 30 - 40 - 50 - 60 - 70 - 80 - 90 - 100 - 110 - 120 - 130
140 - 150 - 160 - 170 - 180 - 190 - 200 - 210 - 220 - 230 - 240 - 250 - 260
270 - 280 - 290 - 300 - 310 - 320 - 330 - 340 - 350 - 360 - 370 - 380 - 390
400 - 410 - 420 - 430 - 440 - 450 - 460 - 470 - 480 - 490 - 500 - 510 - 520
530 - 540 - 550 - 560 - 570 - 580 - 590 - 600 - 610 - 620 - 630 - 640 - 650
660 - 670 - 680 - 690 - 700 - 710 - 720 - 730 - 740 - 750 - 760 - 770 - 780
790 - 800 - 810 - 820 - 830 - 840 - 850 - 860 - 870 - 880 - 890 - 900 - 910
920 - 930 - 940 - 950 - 960 - 970 - 980 - 990 - 1000

Notes: ____

____

____

____

Reg Reps: 10 - 20 - 30 - 40 - 50 - 60 - 70 - 80 - 90 - 100 - 110 - 120 - 130
140 - 150 - 160 - 170 - 180 - 190 - 200 - 210 - 220 - 230 - 240 - 250 - 260
270 - 280 - 290 - 300 - 310 - 320 - 330 - 340 - 350 - 360 - 370 - 380 - 390
400 - 410 - 420 - 430 - 440 - 450 - 460 - 470 - 480 - 490 - 500 - 510 - 520
530 - 540 - 550 - 560 - 570 - 580 - 590 - 600 - 610 - 620 - 630 - 640 - 650
660 - 670 - 680 - 690 - 700 - 710 - 720 - 730 - 740 - 750 - 760 - 770 - 780
790 - 800 - 810 - 820 - 830 - 840 - 850 - 860 - 870 - 880 - 890 - 900 - 910
920 - 930 - 940 - 950 - 960 - 970 - 980 - 990 - 1000

Notes: ____

____

____

____

Reg Reps: 10 - 20 - 30 - 40 - 50 - 60 - 70 - 80 - 90 - 100 - 110 - 120 - 130
140 - 150 - 160 - 170 - 180 - 190 - 200 - 210 - 220 - 230 - 240 - 250 - 260
270 - 280 - 290 - 300 - 310 - 320 - 330 - 340 - 350 - 360 - 370 - 380 - 390
400 - 410 - 420 - 430 - 440 - 450 - 460 - 470 - 480 - 490 - 500 - 510 - 520
530 - 540 - 550 - 560 - 570 - 580 - 590 - 600 - 610 - 620 - 630 - 640 - 650
660 - 670 - 680 - 690 - 700 - 710 - 720 - 730 - 740 - 750 - 760 - 770 - 780
790 - 800 - 810 - 820 - 830 - 840 - 850 - 860 - 870 - 880 - 890 - 900 - 910
920 - 930 - 940 - 950 - 960 - 970 - 980 - 990 - 1000

Notes: ____

____

____

____

Reg Reps: 10 - 20 - 30 - 40 - 50 - 60 - 70 - 80 - 90 - 100 - 110 - 120 - 130
140 - 150 - 160 - 170 - 180 - 190 - 200 - 210 - 220 - 230 - 240 - 250 - 260
270 - 280 - 290 - 300 - 310 - 320 - 330 - 340 - 350 - 360 - 370 - 380 - 390
400 - 410 - 420 - 430 - 440 - 450 - 460 - 470 - 480 - 490 - 500 - 510 - 520
530 - 540 - 550 - 560 - 570 - 580 - 590 - 600 - 610 - 620 - 630 - 640 - 650
660 - 670 - 680 - 690 - 700 - 710 - 720 - 730 - 740 - 750 - 760 - 770 - 780
790 - 800 - 810 - 820 - 830 - 840 - 850 - 860 - 870 - 880 - 890 - 900 - 910
920 - 930 - 940 - 950 - 960 - 970 - 980 - 990 - 1000

Notes: ____

____

____

____

Reg Reps: 10 - 20 - 30 - 40 - 50 - 60 - 70 - 80 - 90 - 100 - 110 - 120 - 130
140 - 150 - 160 - 170 - 180 - 190 - 200 - 210 - 220 - 230 - 240 - 250 - 260
270 - 280 - 290 - 300 - 310 - 320 - 330 - 340 - 350 - 360 - 370 - 380 - 390
400 - 410 - 420 - 430 - 440 - 450 - 460 - 470 - 480 - 490 - 500 - 510 - 520
530 - 540 - 550 - 560 - 570 - 580 - 590 - 600 - 610 - 620 - 630 - 640 - 650
660 - 670 - 680 - 690 - 700 - 710 - 720 - 730 - 740 - 750 - 760 - 770 - 780
790 - 800 - 810 - 820 - 830 - 840 - 850 - 860 - 870 - 880 - 890 - 900 - 910
920 - 930 - 940 - 950 - 960 - 970 - 980 - 990 - 1000

Notes: ____

____

____

____ **End Date:** ____________

Technique:____________________ **L - R - Alt. Start Date:** _______

Notes:______________________________________

Slow Reps: 10 - 20 - 30 - 40 - 50 - 60 - 70 - 80 - 90 - 100 - 110 - 120 - 130
140 - 150 - 160 - 170 - 180 - 190 - 200 - 210 - 220 - 230 - 240 - 250 - 260
270 - 280 - 290 - 300 - 310 - 320 - 330 - 340 - 350 - 360 - 370 - 380 - 390
400 - 410 - 420 - 430 - 440 - 450 - 460 - 470 - 480 - 490 - 500 - 510 - 520
530 - 540 - 550 - 560 - 570 - 580 - 590 - 600 - 610 - 620 - 630 - 640 - 650
660 - 670 - 680 - 690 - 700 - 710 - 720 - 730 - 740 - 750 - 760 - 770 - 780
790 - 800 - 810 - 820 - 830 - 840 - 850 - 860 - 870 - 880 - 890 - 900 - 910
920 - 930 - 940 - 950 - 960 - 970 - 980 - 990 - 1000

Reg Reps: 10 - 20 - 30 - 40 - 50 - 60 - 70 - 80 - 90 - 100 - 110 - 120 - 130
140 - 150 - 160 - 170 - 180 - 190 - 200 - 210 - 220 - 230 - 240 - 250 - 260
270 - 280 - 290 - 300 - 310 - 320 - 330 - 340 - 350 - 360 - 370 - 380 - 390
400 - 410 - 420 - 430 - 440 - 450 - 460 - 470 - 480 - 490 - 500 - 510 - 520
530 - 540 - 550 - 560 - 570 - 580 - 590 - 600 - 610 - 620 - 630 - 640 - 650
660 - 670 - 680 - 690 - 700 - 710 - 720 - 730 - 740 - 750 - 760 - 770 - 780
790 - 800 - 810 - 820 - 830 - 840 - 850 - 860 - 870 - 880 - 890 - 900 - 910
920 - 930 - 940 - 950 - 960 - 970 - 980 - 990 - 1000

Notes: ______________________________________

Reg Reps: 10 - 20 - 30 - 40 - 50 - 60 - 70 - 80 - 90 - 100 - 110 - 120 - 130
140 - 150 - 160 - 170 - 180 - 190 - 200 - 210 - 220 - 230 - 240 - 250 - 260
270 - 280 - 290 - 300 - 310 - 320 - 330 - 340 - 350 - 360 - 370 - 380 - 390
400 - 410 - 420 - 430 - 440 - 450 - 460 - 470 - 480 - 490 - 500 - 510 - 520
530 - 540 - 550 - 560 - 570 - 580 - 590 - 600 - 610 - 620 - 630 - 640 - 650
660 - 670 - 680 - 690 - 700 - 710 - 720 - 730 - 740 - 750 - 760 - 770 - 780
790 - 800 - 810 - 820 - 830 - 840 - 850 - 860 - 870 - 880 - 890 - 900 - 910
920 - 930 - 940 - 950 - 960 - 970 - 980 - 990 - 1000

Notes: ______________________________________

Reg Reps: 10 - 20 - 30 - 40 - 50 - 60 - 70 - 80 - 90 - 100 - 110 - 120 - 130
140 - 150 - 160 - 170 - 180 - 190 - 200 - 210 - 220 - 230 - 240 - 250 - 260
270 - 280 - 290 - 300 - 310 - 320 - 330 - 340 - 350 - 360 - 370 - 380 - 390
400 - 410 - 420 - 430 - 440 - 450 - 460 - 470 - 480 - 490 - 500 - 510 - 520
530 - 540 - 550 - 560 - 570 - 580 - 590 - 600 - 610 - 620 - 630 - 640 - 650
660 - 670 - 680 - 690 - 700 - 710 - 720 - 730 - 740 - 750 - 760 - 770 - 780
790 - 800 - 810 - 820 - 830 - 840 - 850 - 860 - 870 - 880 - 890 - 900 - 910
920 - 930 - 940 - 950 - 960 - 970 - 980 - 990 - 1000

Notes: ____

____

____

____

Reg Reps: 10 - 20 - 30 - 40 - 50 - 60 - 70 - 80 - 90 - 100 - 110 - 120 - 130
140 - 150 - 160 - 170 - 180 - 190 - 200 - 210 - 220 - 230 - 240 - 250 - 260
270 - 280 - 290 - 300 - 310 - 320 - 330 - 340 - 350 - 360 - 370 - 380 - 390
400 - 410 - 420 - 430 - 440 - 450 - 460 - 470 - 480 - 490 - 500 - 510 - 520
530 - 540 - 550 - 560 - 570 - 580 - 590 - 600 - 610 - 620 - 630 - 640 - 650
660 - 670 - 680 - 690 - 700 - 710 - 720 - 730 - 740 - 750 - 760 - 770 - 780
790 - 800 - 810 - 820 - 830 - 840 - 850 - 860 - 870 - 880 - 890 - 900 - 910
920 - 930 - 940 - 950 - 960 - 970 - 980 - 990 - 1000

Notes: ____

____

____

____

Reg Reps: 10 - 20 - 30 - 40 - 50 - 60 - 70 - 80 - 90 - 100 - 110 - 120 - 130
140 - 150 - 160 - 170 - 180 - 190 - 200 - 210 - 220 - 230 - 240 - 250 - 260
270 - 280 - 290 - 300 - 310 - 320 - 330 - 340 - 350 - 360 - 370 - 380 - 390
400 - 410 - 420 - 430 - 440 - 450 - 460 - 470 - 480 - 490 - 500 - 510 - 520
530 - 540 - 550 - 560 - 570 - 580 - 590 - 600 - 610 - 620 - 630 - 640 - 650
660 - 670 - 680 - 690 - 700 - 710 - 720 - 730 - 740 - 750 - 760 - 770 - 780
790 - 800 - 810 - 820 - 830 - 840 - 850 - 860 - 870 - 880 - 890 - 900 - 910
920 - 930 - 940 - 950 - 960 - 970 - 980 - 990 - 1000

Notes: ____

____

____

______________________________________ **End Date:** __________

Technique:________________________ **L - R - Alt. Start Date:** _______

Notes:____

Slow Reps: 10 - 20 - 30 - 40 - 50 - 60 - 70 - 80 - 90 - 100 - 110 - 120 - 130
140 - 150 - 160 - 170 - 180 - 190 - 200 - 210 - 220 - 230 - 240 - 250 - 260
270 - 280 - 290 - 300 - 310 - 320 - 330 - 340 - 350 - 360 - 370 - 380 - 390
400 - 410 - 420 - 430 - 440 - 450 - 460 - 470 - 480 - 490 - 500 - 510 - 520
530 - 540 - 550 - 560 - 570 - 580 - 590 - 600 - 610 - 620 - 630 - 640 - 650
660 - 670 - 680 - 690 - 700 - 710 - 720 - 730 - 740 - 750 - 760 - 770 - 780
790 - 800 - 810 - 820 - 830 - 840 - 850 - 860 - 870 - 880 - 890 - 900 - 910
920 - 930 - 940 - 950 - 960 - 970 - 980 - 990 - 1000

Reg Reps: 10 - 20 - 30 - 40 - 50 - 60 - 70 - 80 - 90 - 100 - 110 - 120 - 130
140 - 150 - 160 - 170 - 180 - 190 - 200 - 210 - 220 - 230 - 240 - 250 - 260
270 - 280 - 290 - 300 - 310 - 320 - 330 - 340 - 350 - 360 - 370 - 380 - 390
400 - 410 - 420 - 430 - 440 - 450 - 460 - 470 - 480 - 490 - 500 - 510 - 520
530 - 540 - 550 - 560 - 570 - 580 - 590 - 600 - 610 - 620 - 630 - 640 - 650
660 - 670 - 680 - 690 - 700 - 710 - 720 - 730 - 740 - 750 - 760 - 770 - 780
790 - 800 - 810 - 820 - 830 - 840 - 850 - 860 - 870 - 880 - 890 - 900 - 910
920 - 930 - 940 - 950 - 960 - 970 - 980 - 990 - 1000

Notes: ____

Reg Reps: 10 - 20 - 30 - 40 - 50 - 60 - 70 - 80 - 90 - 100 - 110 - 120 - 130
140 - 150 - 160 - 170 - 180 - 190 - 200 - 210 - 220 - 230 - 240 - 250 - 260
270 - 280 - 290 - 300 - 310 - 320 - 330 - 340 - 350 - 360 - 370 - 380 - 390
400 - 410 - 420 - 430 - 440 - 450 - 460 - 470 - 480 - 490 - 500 - 510 - 520
530 - 540 - 550 - 560 - 570 - 580 - 590 - 600 - 610 - 620 - 630 - 640 - 650
660 - 670 - 680 - 690 - 700 - 710 - 720 - 730 - 740 - 750 - 760 - 770 - 780
790 - 800 - 810 - 820 - 830 - 840 - 850 - 860 - 870 - 880 - 890 - 900 - 910
920 - 930 - 940 - 950 - 960 - 970 - 980 - 990 - 1000

Notes: ____

Reg Reps: 10 - 20 - 30 - 40 - 50 - 60 - 70 - 80 - 90 - 100 - 110 - 120 - 130
140 - 150 - 160 - 170 - 180 - 190 - 200 - 210 - 220 - 230 - 240 - 250 - 260
270 - 280 - 290 - 300 - 310 - 320 - 330 - 340 - 350 - 360 - 370 - 380 - 390
400 - 410 - 420 - 430 - 440 - 450 - 460 - 470 - 480 - 490 - 500 - 510 - 520
530 - 540 - 550 - 560 - 570 - 580 - 590 - 600 - 610 - 620 - 630 - 640 - 650
660 - 670 - 680 - 690 - 700 - 710 - 720 - 730 - 740 - 750 - 760 - 770 - 780
790 - 800 - 810 - 820 - 830 - 840 - 850 - 860 - 870 - 880 - 890 - 900 - 910
920 - 930 - 940 - 950 - 960 - 970 - 980 - 990 - 1000

Notes: ____

____

____

____

Reg Reps: 10 - 20 - 30 - 40 - 50 - 60 - 70 - 80 - 90 - 100 - 110 - 120 - 130
140 - 150 - 160 - 170 - 180 - 190 - 200 - 210 - 220 - 230 - 240 - 250 - 260
270 - 280 - 290 - 300 - 310 - 320 - 330 - 340 - 350 - 360 - 370 - 380 - 390
400 - 410 - 420 - 430 - 440 - 450 - 460 - 470 - 480 - 490 - 500 - 510 - 520
530 - 540 - 550 - 560 - 570 - 580 - 590 - 600 - 610 - 620 - 630 - 640 - 650
660 - 670 - 680 - 690 - 700 - 710 - 720 - 730 - 740 - 750 - 760 - 770 - 780
790 - 800 - 810 - 820 - 830 - 840 - 850 - 860 - 870 - 880 - 890 - 900 - 910
920 - 930 - 940 - 950 - 960 - 970 - 980 - 990 - 1000

Notes: ____

____

____

____

Reg Reps: 10 - 20 - 30 - 40 - 50 - 60 - 70 - 80 - 90 - 100 - 110 - 120 - 130
140 - 150 - 160 - 170 - 180 - 190 - 200 - 210 - 220 - 230 - 240 - 250 - 260
270 - 280 - 290 - 300 - 310 - 320 - 330 - 340 - 350 - 360 - 370 - 380 - 390
400 - 410 - 420 - 430 - 440 - 450 - 460 - 470 - 480 - 490 - 500 - 510 - 520
530 - 540 - 550 - 560 - 570 - 580 - 590 - 600 - 610 - 620 - 630 - 640 - 650
660 - 670 - 680 - 690 - 700 - 710 - 720 - 730 - 740 - 750 - 760 - 770 - 780
790 - 800 - 810 - 820 - 830 - 840 - 850 - 860 - 870 - 880 - 890 - 900 - 910
920 - 930 - 940 - 950 - 960 - 970 - 980 - 990 - 1000

Notes: ____

____

____

______________________________ **End Date:** ____________

Technique:________________________ **L - R - Alt. Start Date:** ________

Notes:____

Slow Reps: 10 - 20 - 30 - 40 - 50 - 60 - 70 - 80 - 90 - 100 - 110 - 120 - 130
140 - 150 - 160 - 170 - 180 - 190 - 200 - 210 - 220 - 230 - 240 - 250 - 260
270 - 280 - 290 - 300 - 310 - 320 - 330 - 340 - 350 - 360 - 370 - 380 - 390
400 - 410 - 420 - 430 - 440 - 450 - 460 - 470 - 480 - 490 - 500 - 510 - 520
530 - 540 - 550 - 560 - 570 - 580 - 590 - 600 - 610 - 620 - 630 - 640 - 650
660 - 670 - 680 - 690 - 700 - 710 - 720 - 730 - 740 - 750 - 760 - 770 - 780
790 - 800 - 810 - 820 - 830 - 840 - 850 - 860 - 870 - 880 - 890 - 900 - 910
920 - 930 - 940 - 950 - 960 - 970 - 980 - 990 - 1000

Reg Reps: 10 - 20 - 30 - 40 - 50 - 60 - 70 - 80 - 90 - 100 - 110 - 120 - 130
140 - 150 - 160 - 170 - 180 - 190 - 200 - 210 - 220 - 230 - 240 - 250 - 260
270 - 280 - 290 - 300 - 310 - 320 - 330 - 340 - 350 - 360 - 370 - 380 - 390
400 - 410 - 420 - 430 - 440 - 450 - 460 - 470 - 480 - 490 - 500 - 510 - 520
530 - 540 - 550 - 560 - 570 - 580 - 590 - 600 - 610 - 620 - 630 - 640 - 650
660 - 670 - 680 - 690 - 700 - 710 - 720 - 730 - 740 - 750 - 760 - 770 - 780
790 - 800 - 810 - 820 - 830 - 840 - 850 - 860 - 870 - 880 - 890 - 900 - 910
920 - 930 - 940 - 950 - 960 - 970 - 980 - 990 - 1000

Notes: ____

Reg Reps: 10 - 20 - 30 - 40 - 50 - 60 - 70 - 80 - 90 - 100 - 110 - 120 - 130
140 - 150 - 160 - 170 - 180 - 190 - 200 - 210 - 220 - 230 - 240 - 250 - 260
270 - 280 - 290 - 300 - 310 - 320 - 330 - 340 - 350 - 360 - 370 - 380 - 390
400 - 410 - 420 - 430 - 440 - 450 - 460 - 470 - 480 - 490 - 500 - 510 - 520
530 - 540 - 550 - 560 - 570 - 580 - 590 - 600 - 610 - 620 - 630 - 640 - 650
660 - 670 - 680 - 690 - 700 - 710 - 720 - 730 - 740 - 750 - 760 - 770 - 780
790 - 800 - 810 - 820 - 830 - 840 - 850 - 860 - 870 - 880 - 890 - 900 - 910
920 - 930 - 940 - 950 - 960 - 970 - 980 - 990 - 1000

Notes: ____

Reg Reps: 10 - 20 - 30 - 40 - 50 - 60 - 70 - 80 - 90 - 100 - 110 - 120 - 130
140 - 150 - 160 - 170 - 180 - 190 - 200 - 210 - 220 - 230 - 240 - 250 - 260
270 - 280 - 290 - 300 - 310 - 320 - 330 - 340 - 350 - 360 - 370 - 380 - 390
400 - 410 - 420 - 430 - 440 - 450 - 460 - 470 - 480 - 490 - 500 - 510 - 520
530 - 540 - 550 - 560 - 570 - 580 - 590 - 600 - 610 - 620 - 630 - 640 - 650
660 - 670 - 680 - 690 - 700 - 710 - 720 - 730 - 740 - 750 - 760 - 770 - 780
790 - 800 - 810 - 820 - 830 - 840 - 850 - 860 - 870 - 880 - 890 - 900 - 910
920 - 930 - 940 - 950 - 960 - 970 - 980 - 990 - 1000

Notes: ______________________________

____

____

____

Reg Reps: 10 - 20 - 30 - 40 - 50 - 60 - 70 - 80 - 90 - 100 - 110 - 120 - 130
140 - 150 - 160 - 170 - 180 - 190 - 200 - 210 - 220 - 230 - 240 - 250 - 260
270 - 280 - 290 - 300 - 310 - 320 - 330 - 340 - 350 - 360 - 370 - 380 - 390
400 - 410 - 420 - 430 - 440 - 450 - 460 - 470 - 480 - 490 - 500 - 510 - 520
530 - 540 - 550 - 560 - 570 - 580 - 590 - 600 - 610 - 620 - 630 - 640 - 650
660 - 670 - 680 - 690 - 700 - 710 - 720 - 730 - 740 - 750 - 760 - 770 - 780
790 - 800 - 810 - 820 - 830 - 840 - 850 - 860 - 870 - 880 - 890 - 900 - 910
920 - 930 - 940 - 950 - 960 - 970 - 980 - 990 - 1000

Notes: ______________________________

____

____

____

Reg Reps: 10 - 20 - 30 - 40 - 50 - 60 - 70 - 80 - 90 - 100 - 110 - 120 - 130
140 - 150 - 160 - 170 - 180 - 190 - 200 - 210 - 220 - 230 - 240 - 250 - 260
270 - 280 - 290 - 300 - 310 - 320 - 330 - 340 - 350 - 360 - 370 - 380 - 390
400 - 410 - 420 - 430 - 440 - 450 - 460 - 470 - 480 - 490 - 500 - 510 - 520
530 - 540 - 550 - 560 - 570 - 580 - 590 - 600 - 610 - 620 - 630 - 640 - 650
660 - 670 - 680 - 690 - 700 - 710 - 720 - 730 - 740 - 750 - 760 - 770 - 780
790 - 800 - 810 - 820 - 830 - 840 - 850 - 860 - 870 - 880 - 890 - 900 - 910
920 - 930 - 940 - 950 - 960 - 970 - 980 - 990 - 1000

Notes: ______________________________

____

____

______________________________ **End Date:** ____________

Technique:______________________ **L - R - Alt. Start Date:** _______
Notes:____
____
____
____

Slow Reps: 10 - 20 - 30 - 40 - 50 - 60 - 70 - 80 - 90 - 100 - 110 - 120 - 130
140 - 150 - 160 - 170 - 180 - 190 - 200 - 210 - 220 - 230 - 240 - 250 - 260
270 - 280 - 290 - 300 - 310 - 320 - 330 - 340 - 350 - 360 - 370 - 380 - 390
400 - 410 - 420 - 430 - 440 - 450 - 460 - 470 - 480 - 490 - 500 - 510 - 520
530 - 540 - 550 - 560 - 570 - 580 - 590 - 600 - 610 - 620 - 630 - 640 - 650
660 - 670 - 680 - 690 - 700 - 710 - 720 - 730 - 740 - 750 - 760 - 770 - 780
790 - 800 - 810 - 820 - 830 - 840 - 850 - 860 - 870 - 880 - 890 - 900 - 910
920 - 930 - 940 - 950 - 960 - 970 - 980 - 990 - 1000

Reg Reps: 10 - 20 - 30 - 40 - 50 - 60 - 70 - 80 - 90 - 100 - 110 - 120 - 130
140 - 150 - 160 - 170 - 180 - 190 - 200 - 210 - 220 - 230 - 240 - 250 - 260
270 - 280 - 290 - 300 - 310 - 320 - 330 - 340 - 350 - 360 - 370 - 380 - 390
400 - 410 - 420 - 430 - 440 - 450 - 460 - 470 - 480 - 490 - 500 - 510 - 520
530 - 540 - 550 - 560 - 570 - 580 - 590 - 600 - 610 - 620 - 630 - 640 - 650
660 - 670 - 680 - 690 - 700 - 710 - 720 - 730 - 740 - 750 - 760 - 770 - 780
790 - 800 - 810 - 820 - 830 - 840 - 850 - 860 - 870 - 880 - 890 - 900 - 910
920 - 930 - 940 - 950 - 960 - 970 - 980 - 990 - 1000
Notes: ____
____
____
____

Reg Reps: 10 - 20 - 30 - 40 - 50 - 60 - 70 - 80 - 90 - 100 - 110 - 120 - 130
140 - 150 - 160 - 170 - 180 - 190 - 200 - 210 - 220 - 230 - 240 - 250 - 260
270 - 280 - 290 - 300 - 310 - 320 - 330 - 340 - 350 - 360 - 370 - 380 - 390
400 - 410 - 420 - 430 - 440 - 450 - 460 - 470 - 480 - 490 - 500 - 510 - 520
530 - 540 - 550 - 560 - 570 - 580 - 590 - 600 - 610 - 620 - 630 - 640 - 650
660 - 670 - 680 - 690 - 700 - 710 - 720 - 730 - 740 - 750 - 760 - 770 - 780
790 - 800 - 810 - 820 - 830 - 840 - 850 - 860 - 870 - 880 - 890 - 900 - 910
920 - 930 - 940 - 950 - 960 - 970 - 980 - 990 - 1000
Notes: ____
____
____
____

Reg Reps: 10 - 20 - 30 - 40 - 50 - 60 - 70 - 80 - 90 - 100 - 110 - 120 - 130 - 140 - 150 - 160 - 170 - 180 - 190 - 200 - 210 - 220 - 230 - 240 - 250 - 260 - 270 - 280 - 290 - 300 - 310 - 320 - 330 - 340 - 350 - 360 - 370 - 380 - 390 - 400 - 410 - 420 - 430 - 440 - 450 - 460 - 470 - 480 - 490 - 500 - 510 - 520 - 530 - 540 - 550 - 560 - 570 - 580 - 590 - 600 - 610 - 620 - 630 - 640 - 650 - 660 - 670 - 680 - 690 - 700 - 710 - 720 - 730 - 740 - 750 - 760 - 770 - 780 - 790 - 800 - 810 - 820 - 830 - 840 - 850 - 860 - 870 - 880 - 890 - 900 - 910 - 920 - 930 - 940 - 950 - 960 - 970 - 980 - 990 - 1000

Notes: ____

____

____

____

Reg Reps: 10 - 20 - 30 - 40 - 50 - 60 - 70 - 80 - 90 - 100 - 110 - 120 - 130 - 140 - 150 - 160 - 170 - 180 - 190 - 200 - 210 - 220 - 230 - 240 - 250 - 260 - 270 - 280 - 290 - 300 - 310 - 320 - 330 - 340 - 350 - 360 - 370 - 380 - 390 - 400 - 410 - 420 - 430 - 440 - 450 - 460 - 470 - 480 - 490 - 500 - 510 - 520 - 530 - 540 - 550 - 560 - 570 - 580 - 590 - 600 - 610 - 620 - 630 - 640 - 650 - 660 - 670 - 680 - 690 - 700 - 710 - 720 - 730 - 740 - 750 - 760 - 770 - 780 - 790 - 800 - 810 - 820 - 830 - 840 - 850 - 860 - 870 - 880 - 890 - 900 - 910 - 920 - 930 - 940 - 950 - 960 - 970 - 980 - 990 - 1000

Notes: ____

____

____

____

Reg Reps: 10 - 20 - 30 - 40 - 50 - 60 - 70 - 80 - 90 - 100 - 110 - 120 - 130 - 140 - 150 - 160 - 170 - 180 - 190 - 200 - 210 - 220 - 230 - 240 - 250 - 260 - 270 - 280 - 290 - 300 - 310 - 320 - 330 - 340 - 350 - 360 - 370 - 380 - 390 - 400 - 410 - 420 - 430 - 440 - 450 - 460 - 470 - 480 - 490 - 500 - 510 - 520 - 530 - 540 - 550 - 560 - 570 - 580 - 590 - 600 - 610 - 620 - 630 - 640 - 650 - 660 - 670 - 680 - 690 - 700 - 710 - 720 - 730 - 740 - 750 - 760 - 770 - 780 - 790 - 800 - 810 - 820 - 830 - 840 - 850 - 860 - 870 - 880 - 890 - 900 - 910 - 920 - 930 - 940 - 950 - 960 - 970 - 980 - 990 - 1000

Notes: ____

____

____

______________________________ **End Date:** ____________

Technique:________________________ **L - R - Alt. Start Date:** ________
Notes:____

Slow Reps: 10 - 20 - 30 - 40 - 50 - 60 - 70 - 80 - 90 - 100 - 110 - 120 - 130
140 - 150 - 160 - 170 - 180 - 190 - 200 - 210 - 220 - 230 - 240 - 250 - 260
270 - 280 - 290 - 300 - 310 - 320 - 330 - 340 - 350 - 360 - 370 - 380 - 390
400 - 410 - 420 - 430 - 440 - 450 - 460 - 470 - 480 - 490 - 500 - 510 - 520
530 - 540 - 550 - 560 - 570 - 580 - 590 - 600 - 610 - 620 - 630 - 640 - 650
660 - 670 - 680 - 690 - 700 - 710 - 720 - 730 - 740 - 750 - 760 - 770 - 780
790 - 800 - 810 - 820 - 830 - 840 - 850 - 860 - 870 - 880 - 890 - 900 - 910
920 - 930 - 940 - 950 - 960 - 970 - 980 - 990 - 1000

Reg Reps: 10 - 20 - 30 - 40 - 50 - 60 - 70 - 80 - 90 - 100 - 110 - 120 - 130
140 - 150 - 160 - 170 - 180 - 190 - 200 - 210 - 220 - 230 - 240 - 250 - 260
270 - 280 - 290 - 300 - 310 - 320 - 330 - 340 - 350 - 360 - 370 - 380 - 390
400 - 410 - 420 - 430 - 440 - 450 - 460 - 470 - 480 - 490 - 500 - 510 - 520
530 - 540 - 550 - 560 - 570 - 580 - 590 - 600 - 610 - 620 - 630 - 640 - 650
660 - 670 - 680 - 690 - 700 - 710 - 720 - 730 - 740 - 750 - 760 - 770 - 780
790 - 800 - 810 - 820 - 830 - 840 - 850 - 860 - 870 - 880 - 890 - 900 - 910
920 - 930 - 940 - 950 - 960 - 970 - 980 - 990 - 1000
Notes: ____

Reg Reps: 10 - 20 - 30 - 40 - 50 - 60 - 70 - 80 - 90 - 100 - 110 - 120 - 130
140 - 150 - 160 - 170 - 180 - 190 - 200 - 210 - 220 - 230 - 240 - 250 - 260
270 - 280 - 290 - 300 - 310 - 320 - 330 - 340 - 350 - 360 - 370 - 380 - 390
400 - 410 - 420 - 430 - 440 - 450 - 460 - 470 - 480 - 490 - 500 - 510 - 520
530 - 540 - 550 - 560 - 570 - 580 - 590 - 600 - 610 - 620 - 630 - 640 - 650
660 - 670 - 680 - 690 - 700 - 710 - 720 - 730 - 740 - 750 - 760 - 770 - 780
790 - 800 - 810 - 820 - 830 - 840 - 850 - 860 - 870 - 880 - 890 - 900 - 910
920 - 930 - 940 - 950 - 960 - 970 - 980 - 990 - 1000
Notes: ____

Reg Reps: 10 - 20 - 30 - 40 - 50 - 60 - 70 - 80 - 90 - 100 - 110 - 120 - 130
140 - 150 - 160 - 170 - 180 - 190 - 200 - 210 - 220 - 230 - 240 - 250 - 260
270 - 280 - 290 - 300 - 310 - 320 - 330 - 340 - 350 - 360 - 370 - 380 - 390
400 - 410 - 420 - 430 - 440 - 450 - 460 - 470 - 480 - 490 - 500 - 510 - 520
530 - 540 - 550 - 560 - 570 - 580 - 590 - 600 - 610 - 620 - 630 - 640 - 650
660 - 670 - 680 - 690 - 700 - 710 - 720 - 730 - 740 - 750 - 760 - 770 - 780
790 - 800 - 810 - 820 - 830 - 840 - 850 - 860 - 870 - 880 - 890 - 900 - 910
920 - 930 - 940 - 950 - 960 - 970 - 980 - 990 - 1000

Notes: ____

____

____

____

Reg Reps: 10 - 20 - 30 - 40 - 50 - 60 - 70 - 80 - 90 - 100 - 110 - 120 - 130
140 - 150 - 160 - 170 - 180 - 190 - 200 - 210 - 220 - 230 - 240 - 250 - 260
270 - 280 - 290 - 300 - 310 - 320 - 330 - 340 - 350 - 360 - 370 - 380 - 390
400 - 410 - 420 - 430 - 440 - 450 - 460 - 470 - 480 - 490 - 500 - 510 - 520
530 - 540 - 550 - 560 - 570 - 580 - 590 - 600 - 610 - 620 - 630 - 640 - 650
660 - 670 - 680 - 690 - 700 - 710 - 720 - 730 - 740 - 750 - 760 - 770 - 780
790 - 800 - 810 - 820 - 830 - 840 - 850 - 860 - 870 - 880 - 890 - 900 - 910
920 - 930 - 940 - 950 - 960 - 970 - 980 - 990 - 1000

Notes: ____

____

____

____

Reg Reps: 10 - 20 - 30 - 40 - 50 - 60 - 70 - 80 - 90 - 100 - 110 - 120 - 130
140 - 150 - 160 - 170 - 180 - 190 - 200 - 210 - 220 - 230 - 240 - 250 - 260
270 - 280 - 290 - 300 - 310 - 320 - 330 - 340 - 350 - 360 - 370 - 380 - 390
400 - 410 - 420 - 430 - 440 - 450 - 460 - 470 - 480 - 490 - 500 - 510 - 520
530 - 540 - 550 - 560 - 570 - 580 - 590 - 600 - 610 - 620 - 630 - 640 - 650
660 - 670 - 680 - 690 - 700 - 710 - 720 - 730 - 740 - 750 - 760 - 770 - 780
790 - 800 - 810 - 820 - 830 - 840 - 850 - 860 - 870 - 880 - 890 - 900 - 910
920 - 930 - 940 - 950 - 960 - 970 - 980 - 990 - 1000

Notes: ____

____

____

________________________________ **End Date:** __________

Technique:____________________ **L - R - Alt. Start Date:** _______

Notes:____

Slow Reps: 10 - 20 - 30 - 40 - 50 - 60 - 70 - 80 - 90 - 100 - 110 - 120 - 130
140 - 150 - 160 - 170 - 180 - 190 - 200 - 210 - 220 - 230 - 240 - 250 - 260
270 - 280 - 290 - 300 - 310 - 320 - 330 - 340 - 350 - 360 - 370 - 380 - 390
400 - 410 - 420 - 430 - 440 - 450 - 460 - 470 - 480 - 490 - 500 - 510 - 520
530 - 540 - 550 - 560 - 570 - 580 - 590 - 600 - 610 - 620 - 630 - 640 - 650
660 - 670 - 680 - 690 - 700 - 710 - 720 - 730 - 740 - 750 - 760 - 770 - 780
790 - 800 - 810 - 820 - 830 - 840 - 850 - 860 - 870 - 880 - 890 - 900 - 910
920 - 930 - 940 - 950 - 960 - 970 - 980 - 990 - 1000

Reg Reps: 10 - 20 - 30 - 40 - 50 - 60 - 70 - 80 - 90 - 100 - 110 - 120 - 130
140 - 150 - 160 - 170 - 180 - 190 - 200 - 210 - 220 - 230 - 240 - 250 - 260
270 - 280 - 290 - 300 - 310 - 320 - 330 - 340 - 350 - 360 - 370 - 380 - 390
400 - 410 - 420 - 430 - 440 - 450 - 460 - 470 - 480 - 490 - 500 - 510 - 520
530 - 540 - 550 - 560 - 570 - 580 - 590 - 600 - 610 - 620 - 630 - 640 - 650
660 - 670 - 680 - 690 - 700 - 710 - 720 - 730 - 740 - 750 - 760 - 770 - 780
790 - 800 - 810 - 820 - 830 - 840 - 850 - 860 - 870 - 880 - 890 - 900 - 910
920 - 930 - 940 - 950 - 960 - 970 - 980 - 990 - 1000

Notes: ____

Reg Reps: 10 - 20 - 30 - 40 - 50 - 60 - 70 - 80 - 90 - 100 - 110 - 120 - 130
140 - 150 - 160 - 170 - 180 - 190 - 200 - 210 - 220 - 230 - 240 - 250 - 260
270 - 280 - 290 - 300 - 310 - 320 - 330 - 340 - 350 - 360 - 370 - 380 - 390
400 - 410 - 420 - 430 - 440 - 450 - 460 - 470 - 480 - 490 - 500 - 510 - 520
530 - 540 - 550 - 560 - 570 - 580 - 590 - 600 - 610 - 620 - 630 - 640 - 650
660 - 670 - 680 - 690 - 700 - 710 - 720 - 730 - 740 - 750 - 760 - 770 - 780
790 - 800 - 810 - 820 - 830 - 840 - 850 - 860 - 870 - 880 - 890 - 900 - 910
920 - 930 - 940 - 950 - 960 - 970 - 980 - 990 - 1000

Notes: ____

Reg Reps: 10 - 20 - 30 - 40 - 50 - 60 - 70 - 80 - 90 - 100 - 110 - 120 - 130
140 - 150 - 160 - 170 - 180 - 190 - 200 - 210 - 220 - 230 - 240 - 250 - 260
270 - 280 - 290 - 300 - 310 - 320 - 330 - 340 - 350 - 360 - 370 - 380 - 390
400 - 410 - 420 - 430 - 440 - 450 - 460 - 470 - 480 - 490 - 500 - 510 - 520
530 - 540 - 550 - 560 - 570 - 580 - 590 - 600 - 610 - 620 - 630 - 640 - 650
660 - 670 - 680 - 690 - 700 - 710 - 720 - 730 - 740 - 750 - 760 - 770 - 780
790 - 800 - 810 - 820 - 830 - 840 - 850 - 860 - 870 - 880 - 890 - 900 - 910
920 - 930 - 940 - 950 - 960 - 970 - 980 - 990 - 1000

Notes: ____

____

____

____

Reg Reps: 10 - 20 - 30 - 40 - 50 - 60 - 70 - 80 - 90 - 100 - 110 - 120 - 130
140 - 150 - 160 - 170 - 180 - 190 - 200 - 210 - 220 - 230 - 240 - 250 - 260
270 - 280 - 290 - 300 - 310 - 320 - 330 - 340 - 350 - 360 - 370 - 380 - 390
400 - 410 - 420 - 430 - 440 - 450 - 460 - 470 - 480 - 490 - 500 - 510 - 520
530 - 540 - 550 - 560 - 570 - 580 - 590 - 600 - 610 - 620 - 630 - 640 - 650
660 - 670 - 680 - 690 - 700 - 710 - 720 - 730 - 740 - 750 - 760 - 770 - 780
790 - 800 - 810 - 820 - 830 - 840 - 850 - 860 - 870 - 880 - 890 - 900 - 910
920 - 930 - 940 - 950 - 960 - 970 - 980 - 990 - 1000

Notes: ____

____

____

____

Reg Reps: 10 - 20 - 30 - 40 - 50 - 60 - 70 - 80 - 90 - 100 - 110 - 120 - 130
140 - 150 - 160 - 170 - 180 - 190 - 200 - 210 - 220 - 230 - 240 - 250 - 260
270 - 280 - 290 - 300 - 310 - 320 - 330 - 340 - 350 - 360 - 370 - 380 - 390
400 - 410 - 420 - 430 - 440 - 450 - 460 - 470 - 480 - 490 - 500 - 510 - 520
530 - 540 - 550 - 560 - 570 - 580 - 590 - 600 - 610 - 620 - 630 - 640 - 650
660 - 670 - 680 - 690 - 700 - 710 - 720 - 730 - 740 - 750 - 760 - 770 - 780
790 - 800 - 810 - 820 - 830 - 840 - 850 - 860 - 870 - 880 - 890 - 900 - 910
920 - 930 - 940 - 950 - 960 - 970 - 980 - 990 - 1000

Notes: ____

____

____

______________________________ **End Date:** ____________

ADDITIONAL NOTES

Previous Page:_______ Technique:___________________________ Rep #:_____

Previous Page:_______ Technique:___________________________ Rep #:_____

Previous Page:_______ Technique:___________________________ Rep #:_____

Previous Page:_______ Technique:___________________________ Rep #:_____

ADDITIONAL NOTES

Previous Page:_______ Technique:___________________________ Rep #:_____

Previous Page:_______ Technique:___________________________ Rep #:_____

Previous Page:_______ Technique:___________________________ Rep #:_____

Previous Page:_______ Technique:___________________________ Rep #:_____

ADDITIONAL NOTES

Previous Page:_______ Technique:_____________________________ Rep #:_____

Previous Page:_______ Technique:_____________________________ Rep #:_____

Previous Page:_______ Technique:_____________________________ Rep #:_____

Previous Page:_______ Technique:_____________________________ Rep #:_____

ADDITIONAL NOTES

Previous Page:_______ Technique:____________________________ Rep #:_____

Previous Page:_______ Technique:____________________________ Rep #:_____

Previous Page:_______ Technique:____________________________ Rep #:_____

Previous Page:_______ Technique:____________________________ Rep #:_____

ADDITIONAL NOTES

Previous Page:_______ Technique:____________________________ Rep #:_____

Previous Page:_______ Technique:____________________________ Rep #:_____

Previous Page:_______ Technique:____________________________ Rep #:_____

Previous Page:_______ Technique:____________________________ Rep #:_____

ADDITIONAL NOTES

Previous Page:_______ Technique:____________________________ Rep #:_____

Previous Page:_______ Technique:____________________________ Rep #:_____

Previous Page:_______ Technique:____________________________ Rep #:_____

Previous Page:_______ Technique:____________________________ Rep #:_____

ADDITIONAL NOTES

Previous Page:_______ Technique:_____________________________ Rep #:_____

Previous Page:_______ Technique:_____________________________ Rep #:_____

Previous Page:_______ Technique:_____________________________ Rep #:_____

Previous Page:_______ Technique:_____________________________ Rep #:_____

ADDITIONAL NOTES

Previous Page:_______ Technique:_____________________________ Rep #:_____

Previous Page:_______ Technique:_____________________________ Rep #:_____

Previous Page:_______ Technique:_____________________________ Rep #:_____

Previous Page:_______ Technique:_____________________________ Rep #:_____

ADDITIONAL NOTES

Previous Page:_______ Technique:___________________________ Rep #:_____

Previous Page:_______ Technique:___________________________ Rep #:_____

Previous Page:_______ Technique:___________________________ Rep #:_____

Previous Page:_______ Technique:___________________________ Rep #:_____

ADDITIONAL NOTES

Previous Page:_______ Technique:____________________________ Rep #:_____

Previous Page:_______ Technique:____________________________ Rep #:_____

Previous Page:_______ Technique:____________________________ Rep #:_____

Previous Page:_______ Technique:____________________________ Rep #:_____

ADDITIONAL NOTES

Previous Page:_______ Technique:_____________________________ Rep #:_____

Previous Page:_______ Technique:_____________________________ Rep #:_____

Previous Page:_______ Technique:_____________________________ Rep #:_____

Previous Page:_______ Technique:_____________________________ Rep #:_____

ADDITIONAL NOTES

Previous Page:_______ Technique:___________________________ Rep #:_____

Previous Page:_______ Technique:___________________________ Rep #:_____

Previous Page:_______ Technique:___________________________ Rep #:_____

Previous Page:_______ Technique:___________________________ Rep #:_____

ADDITIONAL NOTES

Previous Page:_______ Technique:___________________________ Rep #:_____

Previous Page:_______ Technique:___________________________ Rep #:_____

Previous Page:_______ Technique:___________________________ Rep #:_____

Previous Page:_______ Technique:___________________________ Rep #:_____

ADDITIONAL NOTES

Previous Page:_______ Technique:___________________________ Rep #:_____

Previous Page:_______ Technique:___________________________ Rep #:_____

Previous Page:_______ Technique:___________________________ Rep #:_____

Previous Page:_______ Technique:___________________________ Rep #:_____

ADDITIONAL NOTES

Previous Page:_______ Technique:_____________________________ Rep #:_____

Previous Page:_______ Technique:_____________________________ Rep #:_____

Previous Page:_______ Technique:_____________________________ Rep #:_____

Previous Page:_______ Technique:_____________________________ Rep #:_____

ADDITIONAL NOTES

Previous Page:_______ Technique:____________________________ Rep #:_____

Previous Page:_______ Technique:____________________________ Rep #:_____

Previous Page:_______ Technique:____________________________ Rep #:_____

Previous Page:_______ Technique:____________________________ Rep #:_____

Made in United States
Troutdale, OR
11/03/2026

56844438R00073